AFFAN JAVED

Reality Is Not What You Think

How Modern Physics Rewrites Space, Time, Matter, and the Universe

First edition

This book was professionally typeset on Reedsy.
Find out more at reedsy.com

Contents

Preface

There are moments when the universe quietly refuses to behave the way we expect. A clock ticks differently when it moves. A particle passes through two places at once. Empty space weighs something. The cosmos expands not just outward, but faster and faster, as if driven by a hidden hand. These are not philosophical riddles or speculative fantasies. They are measured facts, confirmed by experiments so precise that they underpin the technologies we rely on every day. And yet, taken together, they tell us something profoundly unsettling: reality is not what common sense says it should be.

This book begins with a simple premise that the modern picture of the universe is stranger, more beautiful, and more humbling than any myth we have ever told. For most of human history, we believed the world was made of solid things moving through a fixed stage of space and time. Cause followed effect in a neat, predictable order. If we knew enough, we thought, we could in principle predict everything. Physics in the twentieth and twenty-first centuries shattered that vision. What replaced it was not chaos, but a deeper order that resists intuition and demands imagination.

At the smallest scales, reality behaves like a riddle that answers itself differently depending on how it is asked. Matter is not merely "stuff," and light is not merely a wave or a particle. Observation is not passive. Uncertainty is not a flaw in our instruments but a fundamental feature of nature. At the largest scales, space and time are no longer a rigid backdrop but active participants in the cosmic drama, stretching, bending, and evolving. Gravity is not a force in the traditional sense, and time itself does not flow at the same rate for everyone.

This book is organized as a journey through these ideas, not as a technical manual but as a guided exploration. Each chapter focuses on a core discovery or concept that changed how we understand the universe. The goal is not to overwhelm with equations or jargon, but to reveal the logic, evidence, and wonder behind the ideas. You will see how a single experiment with light can undermine centuries of certainty, how a thought experiment about elevators leads to a new theory of gravity, and how faint signals from distant galaxies allow us to reconstruct the history of the cosmos.

The structure follows a natural progression. We begin with the quantum world, where the rules of everyday experience fail most dramatically. We then step outward into relativity and the structure of space-time, where motion, energy, and gravity are woven into a single framework. From there, we widen our view to the universe as a whole its origin, its composition, and its fate. Finally, we ask the unavoidable question: what does all of this mean, not just for physics, but for technology, knowledge, and our place in the universe?

You do not need a background in physics to read this book. What you need is curiosity and a willingness to let go of the idea that reality must conform to intuition. The universe does not owe us simplicity. Yet again and again, when we follow the evidence honestly, we find patterns of astonishing coherence beneath the strangeness.

If there is a single message running through these pages, it is this: the universe is not only stranger than we imagine, it is stranger than we can imagine and still, somehow, understandable. That tension between mystery and comprehension is the heartbeat of modern physics. It is also an invitation. This book invites you to step into that tension, to see how far human thought has reached, and to glimpse how much remains unknown.

The story that follows is not finished. It is a living narrative, shaped by experiments still being built and questions still unanswered. But even now, it reveals a universe more subtle, more interconnected, and more awe-inspiring than the one we once thought we inhabited.

Introduction

Introduction: When the Universe Stopped Making Sense

For thousands of years, humans believed they understood reality. The ground beneath our feet was solid and dependable. Time flowed steadily forward, ticking at the same rate for everyone, everywhere in the universe. Objects had definite positions and velocities, following predictable paths through space. Cause always preceded effect. If you knew the present state of any system with enough precision, you could calculate its future with absolute certainty. The universe operated like an enormous, magnificent clockwork mechanism, perhaps complex and intricate, but ultimately knowable, governed by mathematical rules we could discover and master through careful observation and experimentation.

Then, in the early twentieth century, everything changed.

Physicists began probing the extreme edges of reality. They examined the incredibly small world of atoms and subatomic particles. They studied phenomena occurring at incredibly high speeds, approaching the cosmic speed limit of light itself. They investigated the incredibly massive domains of stars, galaxies, and the entire cosmos. What they discovered in these extreme regimes shattered every intuitive notion humans had ever held about the fundamental nature of existence.

They found particles that could exist in multiple places simultaneously until observed. They discovered that time itself slows down for objects moving at high velocities, and that two observers could legitimately disagree about whether events happened simultaneously or in sequence. They revealed that empty space was not empty at all but seething with quantum energy and virtual particles constantly popping in and out of existence. They demonstrated that objects separated by vast distances across the universe could remain mysteriously connected, somehow influencing each other's properties instantaneously despite nothing traveling between them. They proved that matter and energy were not separate categories but different manifestations of the same underlying reality, convertible from one form to the other. They showed that the entire universe had a definite beginning in time and continues to expand outward into something that cannot be called space in any conventional sense.

These were not philosophical musings or abstract mathematical games played by ivory tower academics. These were concrete experimental facts, confirmed by thousands upon thousands of precise measurements, replicated in laboratories around the world, forming the theoretical foundation of technologies you interact with every single day. The smartphone in your pocket relies on quantum mechanics to operate its transistors and memory chips. The GPS system guiding your car must account for relativistic time dilation or it would accumulate errors of several miles per day. The medical imaging technologies that detect tumors and save lives depend on our understanding of how particles behave in ways that violate classical physics. The nuclear power plants generating electricity harness Einstein's discovery that mass can be converted into energy. Even the humble LED light bulb in your lamp is a quantum device that would be impossible to design without modern physics.

This book tells the story of modern physics, but not as a dry collection of formulas, equations, and abstract mathematical theorems. Instead, it presents this material as it truly is: a revolution in human understanding as profound as the Copernican revelation that Earth orbits the Sun rather than standing at the center of creation, or Darwin's discovery that humans evolved from earlier species rather than being specially created. Modern physics represents a complete overthrow of our intuitive picture of reality, replacing it with something initially bizarre and counter intuitive, yet ultimately more accurate, more powerful, and in many ways more beautiful than what came before.

The journey will take you from the smallest scales imaginable to the largest structures in the cosmos. You will encounter ideas that seem impossible at first, that appear to violate logic itself. You will read about experiments so clever and results so strange that even the physicists who performed them initially refused to believe their own data. You will discover that many things you thought were obviously true about the universe turn out to be approximately correct only in the narrow range of everyday human experience, but fundamentally wrong when examined more carefully.

You do not need advanced mathematics to understand these revolutionary ideas. You do not need a science degree or years of specialized training. You simply need intellectual curiosity and a willingness to question assumptions you have held since childhood. You need to be open to the possibility that reality operates according to rules very different from those our evolutionary history prepared us to expect. Because once you see the universe as modern physics reveals it, once you understand what experiments have actually demonstrated about the nature of space, time, matter, and energy, you can never see the world quite the same way again. Colors you once thought existed in objects reveal themselves as constructions in your brain. Solidity you once thought defined matter turns out to be an illusion created by electromagnetic forces. The flow of time you once thought absolute and universal reveals itself as relative and flexible. The separation between objects you once thought inviolable shows itself permeable through quantum entanglement.

The universe does not care what seems reasonable to creatures who evolved to hunt prey and gather fruit on the African savanna. It does not adjust itself to match human intuitions developed in a world of moderate speeds, moderate energies, and moderate scales. Reality simply is what it is, operating according to its own deep mathematical structures, and our job as curious beings is to discover those structures through observation and experiment, even when they surprise us, even when they contradict cherished beliefs, even when they force us to completely rethink concepts as basic as existence, causation, and identity.

What follows is the story of that discovery. It is a story still being written, because physics has not answered every question. We still do not fully understand how gravity works at quantum scales. We do not know what most of the universe is made of. We do not know whether our universe is the only one or part of an infinite multiverse. We do not know whether space and time are fundamental or emerge from something even more basic. But we have learned enough to know that the universe is far stranger and far more wonderful than previous generations ever imagined.

This introduction serves as an invitation and a warning. An invitation to join one of humanity's greatest intellectual adventures, the quest to understand physical reality at its deepest levels. And a warning that the journey will challenge you, that accepting what experiments actually show about the universe requires abandoning comfortable certainties in favor of uncomfortable truths. If you are ready for that journey, turn the page. The universe is waiting to reveal its secrets, and those secrets are more astonishing than any fiction ever invented.

I

I. The Quantum Strange

The first part of our journey takes us into the quantum realm, the domain of atoms and subatomic particles where reality behaves in ways that seem impossible according to everyday experience. This is where modern physics becomes truly revolutionary, where particles act like waves and waves act like particles, where nothing has definite properties until measured, where objects remain connected across arbitrary distances, and where the very act of observation changes what is observed.

1. The Double Slit Experiment - The Most Beautiful Mystery in Physics

In 1801, Thomas Young performed an experiment that would eventually overthrow everything physicists thought they knew about reality. He cut two narrow parallel slits in an opaque barrier and shined light through them onto a screen positioned behind the barrier. What appeared on the screen was not what anyone expected.

If light consisted of particles, as Isaac Newton had argued, the screen should have shown two bright bands directly behind the two slits, perhaps slightly spread out but basically matching the slit pattern. Instead, Young saw something entirely different: a pattern of multiple bright and dark bands spread across the screen, with the brightest region not behind either slit but midway between them.

This interference pattern was unmistakable evidence that light behaves as a wave. When a wave passes through two slits, the two emerging wavefront spread outward and overlap. Where the peaks of the two waves align, they reinforce each other, creating bright bands. Where a peak from one slit meets a trough from the other, they cancel out, creating dark bands. This is exactly what Young observed. Light, he concluded, must be a wave motion, not a stream of particles.

For over a century, Young's wave theory of light prevailed. Physicists developed sophisticated mathematical descriptions of electromagnetic waves, culminating in James Clerk Maxwell's equations that unified electricity, magnetism, and light into a single theoretical framework. The wave nature of light seemed established beyond any reasonable doubt.

Then in 1905, Albert Einstein showed that light sometimes behaves as if it consists of discrete particles, now called photons. He was analyzing the photoelectric effect, the phenomenon where light striking certain metals ejects electrons from their surface. Experiments showed that dim blue light would eject electrons immediately, but no amount of bright red light, no matter how intense, could eject any electrons at all. This made no sense if light was purely a wave, because brighter light should deliver more energy and eventually dislodge electrons regardless of color.

Einstein explained the photoelectric effect by proposing that light arrives in quantum packets of energy, with each photon carrying energy proportional to the light's frequency. Blue light consists of high energy photons capable of ejecting electrons. Red light consists of low energy photons, and making the light brighter just means more low energy photons arrive, but individually they still lack sufficient energy to free an electron from the metal. This particle picture explained the photoelectric effect perfectly.

So which is it? Is light a wave or a particle? The honest answer, confirmed by countless subsequent experiments, is that light is both and neither. Light exhibits wave properties in some situations and particle properties in others, and no classical concept adequately captures its true nature.

This wave particle duality becomes even stranger when you perform the double slit experiment with matter instead of light. In the 1920s, physicists began exploring whether particles like electrons might also exhibit wave properties. Clinton Davisson and Lester Germer fired electrons at a nickel crystal and observed a diffraction pattern, exactly the sort of pattern waves produce when reflecting from a regular grid. Electrons, fundamental particles of matter that everyone assumed were solid little spheres, were somehow also waves.

But the ultimate demonstration came when physicists repeated Young's double slit experiment using electrons instead of light. They set up a barrier with two narrow slits and fired electrons toward it, with a detection screen positioned behind the barrier to record where each electron arrived.

When both slits were open, the electrons created an interference pattern on the screen, with alternating bands of many electron impacts and regions where almost no electrons arrived. Just like light, the electrons were acting as waves passing through both slits simultaneously, with the two waves interfering to create the characteristic pattern.

When experimenters closed one slit, the interference pattern disappeared and electrons simply piled up in a region directly behind the single open slit, exactly as particles should behave. The interference pattern required both slits to be open.

Here is where the mystery deepens into something truly profound. Experimenters decided to send electrons through the apparatus one at a time, with minutes between individual electrons, ensuring that each electron traversed the apparatus in complete isolation with no other electrons anywhere nearby. Over time, as thousands of individual electrons were detected, the interference pattern gradually built up on the screen, one electron impact at a time.

Think carefully about what this means. Each electron, traveling alone through the apparatus, somehow "knows" that both slits are open and behaves accordingly. Each single electron produces an interference pattern as if it passed through both slits simultaneously and interfered with itself. The interference pattern appears even when electrons are sent through one at a time, absolutely ruling out any explanation based on electrons interacting with each other.

An electron is not splitting into pieces and recombining. When an electron reaches the detection screen, it always arrives whole, creating a single localized impact point, never spread out. Yet the accumulation of many single electron impacts builds up a wave interference pattern that should only be possible if each electron somehow passed through both slits at once.

Physicists naturally wondered what would happen if they tried to observe which slit each electron actually passed through. They added detectors at each slit capable of registering when an electron passed by. The result was shocking: when detectors were active and experimenters could determine which slit each electron used, the interference pattern vanished completely. The electron distribution on the screen now showed two simple peaks, exactly as you would expect for classical particles passing through one slit or the other, never both.

The act of measurement changed the outcome. When no one measured which path the electron took, it exhibited wave behavior and created an interference pattern, as if traveling through both slits simultaneously. When experimenters measured which slit the electron passed through, it exhibited particle behavior and the interference pattern disappeared, as if each electron now committed to traveling through only one slit.

This was not a matter of measurement disturbing the electron through some unintended physical interaction. Experimenters tried ever more subtle detection methods, tried detecting the electrons after they had already passed the slits, tried quantum eraser experiments where the which path information was erased before the final detection, and in every case the conclusion remained the same. When which path information was available, even in principle, the interference pattern vanished. When which path information was not available, even if experimenters chose not to look at detectors that recorded it, the interference pattern appeared.

The universe somehow enforces a trade off between wave-like and particle-like behavior. You can observe interference patterns or you can determine which path a particle took, but you cannot have both simultaneously. Nature does not merely hide this information from you. The information appears not to exist. Before measurement, the electron genuinely does not have a definite position or path. It exists in a quantum superposition of states, a kind of existence that has no classical analogy.

Let me be crystal clear about what experiments have actually demonstrated, because this point is so important and so commonly misunderstood. Before measurement, an electron does not actually pass through the left slit or the right slit while we simply remain ignorant of which one. That explanation, appealing as it sounds, predicts there should be no interference pattern, because each electron really did take only one path and we are merely uncertain about which path. But we do observe an interference pattern, which proves that explanation is wrong.

The electron also does not split into two pieces, with half passing through each slit and recombining afterward. That explanation predicts we should sometimes detect half an electron or see two impacts on the screen from a single electron. We never do. Every electron always arrives whole at exactly one location.

The only explanation consistent with all experimental results is that before measurement, the electron genuinely exists in a superposition of states, having no definite position, not occupying any specific location in space. It is best described mathematically as a wave function, a spread out probability distribution that assigns different probabilities to different possible measurement outcomes. This wave function passes through both slits, interferes with itself, and determines the probability of where the electron will be detected. At the moment of detection, the wave function collapses, the electron localizes to a single position, and it appears whole at one particular spot on the screen.

This is not how anything works in everyday experience. Objects you encounter in daily life have definite positions and follow definite paths even when you are not looking at them. Your car sits in your driveway with a specific location whether or not you go outside to check on it. A baseball traveling toward a bat follows a continuous trajectory, occupying each point along its path in succession.

But at quantum scales, this classical picture fails completely. Particles do not have definite positions or velocities before measurement. They exist in superposition states that defy any visualizable classical description. The double slit experiment demonstrates this as clearly as any experiment can. It shows that the quantum world operates according to rules entirely different from the classical rules that govern everyday objects.

Why do quantum effects vanish for large objects? Why does your car have a definite location while an electron does not? The answer involves decoherence, the process by which quantum super positions become effectively classical through interaction with the environment. A macroscopic object like a car is constantly interacting with its surroundings through countless air molecules bouncing off it, photons reflecting from its surface, heat radiation, gravitational effects, and so forth. These interactions constitute continuous measurement of the car's position, constantly collapsing any potential quantum superposition and forcing the car into a definite classical state.

An electron in a carefully isolated double slit apparatus can maintain quantum coherence because it interacts minimally with its environment during its journey through the slits. The moment it hits the detection screen and transfers energy to atoms in the screen material, that constitutes a measurement and the wave function collapses.

Quantum mechanics provides a precise mathematical framework for calculating the probabilities of different measurement outcomes. Given an initial quantum state, Schrödinger's wave equation determines how the state evolves over time. When a measurement occurs, the Born rule specifies the probability of each possible result. These mathematical tools allow physicists to make predictions of extraordinary accuracy.

But the conceptual interpretation remains deeply puzzling. What is the wave function physically? Does it represent something real existing in space, or merely our knowledge about the system? What constitutes a measurement that collapses the wave function? Does the wave function actually collapse, or do all possible outcomes occur in parallel universes? Physicists have debated these interpretation questions since quantum mechanics was first formulated, and no consensus has emerged.

What is not in question are the experimental facts. The double slit experiment, performed with photons, electrons, atoms, and even large molecules, consistently demonstrates wave particle duality. Particles really do behave as if they pass through multiple locations simultaneously when not measured. Measurement really does change the outcome in ways that cannot be explained by simple disturbance. The quantum world really does operate according to rules that violate classical intuitions about how objects should behave.

This is the foundation of modern physics. Everything else in quantum mechanics builds on the insights revealed by the double slit experiment: that matter exhibits wave like properties, that quantum systems exist in superposition until measured, that measurement outcomes are probabilistic rather than deterministic, and that observation plays a fundamental role in determining physical reality.

The physicist Richard Feynman once said that the double slit experiment contains the only mystery of quantum mechanics, and that understanding this one experiment means understanding all the strange features of the quantum world. While that may be somewhat of an exaggeration, Feynman's point is well taken. If you can accept that electrons genuinely pass through both slits simultaneously when not measured, and that measurement somehow collapses this superposition into a definite outcome, then you have grasped the essential strangeness of quantum mechanics.

Reality at small scales simply does not work the way our everyday experience suggests it should. The universe is not made of tiny solid objects following definite trajectories through space. It is made of quantum entities that exist in superposition, described by wave functions evolving according to Schrödinger's equation, and localizing into definite observed properties only at the moment of measurement. This is not a philosophical speculation or a mathematical convenience. This is what experiments actually show, demonstrated with ever greater precision and sophistication over the past century.

The double slit experiment stands as perhaps the most beautiful and most disturbing demonstration in all of physics. Beautiful because of its elegant simplicity, using only slits, a screen, and a source of quantum particles to reveal the deepest mysteries of nature. Disturbing because it forces us to abandon our classical conception of reality and accept that at its most fundamental level, the universe operates according to quantum principles that defy visualization and challenge our deepest intuitions about existence itself.

Welcome to the quantum world. Nothing here behaves as you expect. And that is just the beginning.

2. Wave or Particle? Yes. - Why Matter Has No Fixed Identity

The double slit experiment revealed that light and matter exhibit both wave properties and particle properties depending on how they are measured. This wave particle duality initially seemed like a paradox, a sign that physics was somehow incomplete or confused. Surely things must be either waves or particles, not both. How can a single entity possess two apparently contradictory natures?

The resolution is that waves and particles are not fundamental categories of reality but rather classical concepts, mental models we developed based on everyday experience. These models work well for describing macroscopic phenomena within certain contexts, but they fail at quantum scales where neither model fully captures what is actually happening.

Consider what we mean by a particle in classical physics. A particle is a localized object occupying a definite position in space at each moment in time. It has a specific velocity, traveling along a continuous trajectory from one location to another. When two particles collide, they interact at a specific point in space and time. You can count particles: there is one particle here, another particle there, two particles in total.

Now consider what we mean by a wave. A wave is a spreading disturbance, extended across a region of space, not localized to any particular point. A wave has wavelength and frequency rather than position and velocity. When two waves meet, they interfere, creating patterns of constructive and destructive interference. You cannot count waves in the same way you count particles: where one wave ends and another begins is often ambiguous.

These classical concepts are incompatible. Something cannot simultaneously have both a definite position and be spread out over space. It cannot simultaneously follow a specific trajectory and exhibit interference effects that require traveling multiple paths. Yet experiments show that quantum entities like electrons and photons exhibit both types of behavior depending on the experimental context.

The key insight that resolved this apparent paradox came from Louis de Broglie in 1924. De Broglie proposed that all matter possesses wave properties, with the wavelength inversely proportional to the momentum. A massive object moving slowly has an extremely short de Broglie wavelength, so short that wave effects are completely negligible and the object behaves as a classical particle. A light, fast moving object like an electron has a longer de Broglie wavelength, long enough that wave effects become observable.

This explained why we observe particle behavior in everyday life but wave behavior in atomic scale experiments. A baseball traveling toward home plate has a de Broglie wavelength around 10 to the negative 34 meters, utterly minuscule compared to the size of atoms. No experiment could possibly detect wave effects at such small scales, so the baseball behaves purely as a particle. An electron traveling at typical speeds in an atom has a de Broglie wavelength comparable to the size of the atom itself, making wave effects dominant and observable.

The wave associated with a quantum particle is not a wave of any physical substance. It is not like a water wave, which is a traveling disturbance in water molecules, or a sound wave, which is a compression wave in air. The quantum wave function is a wave of probability amplitudes. At each point in space, the wave function has a complex number value. The square of the absolute value of this complex number gives the probability density of finding the particle at that location if a position measurement is performed.

This probabilistic interpretation, developed primarily by Max Born, represents a radical departure from classical determinism. In classical physics, if you know the initial position and velocity of a particle precisely, you can calculate its future trajectory with arbitrary accuracy. The uncertainty in prediction arises only from uncertainty in your initial knowledge. In quantum mechanics, even with perfect knowledge of the initial quantum state, you can only calculate probabilities for different measurement outcomes. The outcomes themselves are fundamentally random, not determined by any underlying variables.

Einstein never accepted this. He famously objected that God does not play dice with the universe, suggesting that quantum mechanics must be incomplete, that some deeper theory with hidden variables could restore determinism. But experiments have shown that Einstein was wrong. The universe really is fundamentally probabilistic at quantum scales. The randomness in quantum measurement outcomes is not due to ignorance of hidden variables. It is intrinsic to nature itself.

The wave function describing a quantum particle evolves smoothly and deterministically according to Schrödinger's equation. Given an initial wave function, you can calculate with certainty what the wave function will be at any future time. But the wave function does not directly tell you where the particle is. It tells you only probabilities. When you perform a position measurement, the wave function collapses to a specific outcome chosen randomly according to the probabilities specified by the wave function. After collapse, the particle has a definite measured position, and the wave function now describes a localized state concentrated at that position.

This collapse process remains the most controversial and least understood aspect of quantum mechanics. Schrödinger's equation itself never produces collapse. It describes smooth, continuous, deterministic evolution of wave functions. Collapse happens only during measurement, appearing as a discontinuous jump to a definite outcome. What constitutes a measurement? When exactly does collapse occur? Does the wave function really collapse, or does the universe split into multiple branches, with each possible outcome occurring in a different branch? These questions remain hotly debated, with several competing interpretations of quantum mechanics offering different answers.

Despite this interpretational controversy, the mathematical formalism works perfectly. Physicists can calculate quantum probabilities with extraordinary precision and experiments consistently confirm those predictions. Whatever is really happening during measurement, quantum mechanics gives the right answers.

The wave particle duality extends beyond just electrons and photons. Experiments have demonstrated wave behavior for progressively larger objects: individual atoms, simple molecules, complex molecules, even specially prepared collections of thousands of atoms behaving as a single quantum object. In 1999, researchers performed a double slit experiment with buckminsterfullerene molecules, also called buckyballs, consisting of 60 carbon atoms arranged in a spherical cage. These molecules, containing hundreds of protons, neutrons, and electrons, still exhibited quantum interference, passing through both slits simultaneously despite being macroscopic chemical structures.

The limit to observable quantum behavior is not fundamentally about size but about decoherence. Any interaction between a quantum system and its environment can constitute measurement, collapsing quantum superpositions. For large objects, environmental interactions occur constantly and unavoidably. Air molecules bounce off the object, photons scatter from its surface, gravitational interactions extend to distant masses, heat radiation carries away energy. Each of these interactions entangles the object's quantum state with the environment, effectively measuring the object's properties and destroying quantum coherence.

For small, carefully isolated quantum systems, decoherence can be minimized. Electrons in the double slit experiment remain coherent because they travel through vacuum with minimal environmental interaction. Buckyballs in quantum interference experiments must be heated to release molecules one at a time into a high vacuum and pass through carefully designed slits with minimal interaction. Any environmental interaction strong enough to determine which slit a molecule passed through will destroy the interference pattern.

This explains why you never observe quantum effects in daily life. Your body temperature is 37 degrees Celsius, meaning atoms in your body are in constant thermal motion, colliding with each other trillions of times per second. Photons from surrounding light sources constantly strike your body. Air molecules buffet your skin. These interactions constitute continuous measurement, collapsing any potential quantum superpositions and giving you definite classical properties.

But quantum mechanics is not just relevant at small scales. It is the fundamental theory describing all matter. Classical physics emerges as an approximation valid for large objects in the limit where quantum effects average out. Every solid object around you appears solid only because of quantum mechanics. The electromagnetic forces between atoms that give matter its rigidity and prevent your hand from passing through a table are described by quantum field theory. Chemistry, the science explaining how atoms combine to form molecules, is entirely quantum mechanical. Biology, the study of life, ultimately reduces to quantum mechanical processes occurring in cells.

So is matter fundamentally waves or particles? The honest answer is neither.

Matter is quantum mechanical entities that exhibit wave like behavior in some experimental contexts and particle like behavior in other contexts. When you perform experiments that probe interference effects, you see wave properties. When you perform experiments that measure positions or count discrete impacts, you see particle properties. But neither waves nor particles adequately capture the underlying quantum reality.

Thinking of electrons as tiny solid spheres following definite paths is wrong. Thinking of electrons as spread out waves oscillating through space is also wrong. The electron is a quantum object, described by a wave function living in an abstract mathematical space called Hilbert space. The wave function evolves according to Schrödinger's equation. When you measure the electron, you find it in a definite state with probabilities determined by the wave function. That is the most accurate description we have, and attempts to impose classical intuitions just lead to confusion and contradiction.

Perhaps the most important lesson of wave particle duality is that quantum objects do not have definite properties before measurement. An unmeasured electron does not have a specific position that we just happen not to know. It genuinely lacks a definite position, existing instead in a superposition of many possible positions with various amplitudes. Only measurement forces the electron to take on a definite position, chosen randomly according to quantum probabilities.

This is profoundly different from classical physics, where properties exist independently of observation. A classical particle has a position whether or not anyone measures it. Measurement merely reveals the pre existing property. In quantum mechanics, measurement creates the property. The electron had no definite position before measurement. The measurement outcome is not revealing a hidden pre existing value but rather instantiating a value that did not exist beforehand.

John Bell proved in the 1960s that this is not just an interpretation but a necessary conclusion from quantum mechanics. Bell's theorem shows that no theory based on local hidden variables, no theory where particles have definite properties all along that we simply do not know, can reproduce all quantum predictions. Experiments testing Bell inequalities have confirmed quantum mechanics and ruled out local hidden variable theories. The properties of quantum particles really are indefinite until measured.

This also means that quantum objects lack definite identity in the way classical objects do. If you have two identical electrons, there is no fact of the matter about which electron is which. You cannot label them and track them individually because before measurement they do not occupy definite locations. If you perform an experiment that swaps two identical particles and then swaps them back, you have not merely returned to the original configuration. You have performed a different quantum operation that can produce observable effects through quantum interference.

This quantum indistinguishability of identical particles has major consequences for how matter behaves. It divides all particles into two categories: bosons, which can occupy the same quantum state, and fermions, which obey the Pauli exclusion principle and cannot occupy identical states. Electrons are fermions, which is why atoms have complex shell structure with electrons forced into different orbitals. Photons are bosons, which is why lasers work by having many photons in exactly the same quantum state. The entire structure of the periodic table and all of chemistry emerges from these quantum statistical effects.

Wave particle duality resolves into a deeper understanding: quantum objects are neither waves nor particles but something else entirely, something described beautifully by the mathematics of quantum mechanics but not visualizable in classical terms. When physicists say an electron is a particle, they mean it produces discrete impacts on detectors and can be counted. When they say it is a wave, they mean it produces interference patterns and has wavelength and frequency. Both statements are correct within their respective contexts, but neither captures the complete quantum reality.

Learning to think quantum mechanically requires abandoning classical intuitions. You must stop imagining electrons as tiny billiard balls. You must stop trying to visualize wave functions as physical waves in space. Instead, you must accept the mathematics as the description and understand that quantum objects behave according to quantum rules that have no classical analog.

The reward for this conceptual shift is a far deeper understanding of physical reality. Quantum mechanics explains the stability of atoms, the nature of chemical bonds, the behavior of solids, semiconductors, and superconductors, the operation of lasers and transistors, and countless other phenomena completely inexplicable in classical terms. It provides the foundation for modern technology and reveals that reality at its most fundamental level operates according to principles far stranger and more subtle than everyday experience would ever suggest.

3. The Uncertainty Principle - Nature's Built-In Fuzziness

In 1927, Werner Heisenberg discovered one of the most fundamental limitations in nature. He showed that certain pairs of physical properties cannot simultaneously have definite values, no matter how carefully you design your experiments or how advanced your measurement technology becomes. The more precisely you determine one property, the less precisely the other property can be known. This is not a statement about measurement difficulty or experimental imperfection. It represents an intrinsic feature of quantum reality itself.

The most famous example involves position and momentum. Momentum is mass times velocity, so knowing momentum means knowing how fast something is moving and in which direction. Heisenberg proved that the product of the uncertainty in position and the uncertainty in momentum must always exceed a certain minimum value, specifically Planck's constant divided by four pi. This means if you measure a particle's position with very high precision, pinning it down to a tiny region of space, its momentum becomes correspondingly uncertain, spread over a wide range of possible values. Conversely, if you measure momentum very precisely, the position becomes uncertain, spread over a large region.

This uncertainty principle is not a statement about measurement disturbing the system. Early popular explanations sometimes claimed that measuring position requires hitting the particle with photons, which inevitably disturbs its momentum. While measurement does cause disturbance, Heisenberg's principle goes deeper. The uncertainty exists even before any measurement. A quantum particle simply cannot have both a definite position and a definite momentum simultaneously. These properties are complementary, meaning maximizing knowledge of one minimizes knowledge of the other.

The underlying reason traces back to wave particle duality. Position is a particle property. A localized particle exists at a specific location. Momentum is related to wavelength through de Broglie's relation, so it is a wave property. A wave with definite wavelength extends over space and has no single position. A quantum state with definite momentum is a pure wave extending infinitely through space. A quantum state with definite position is a perfectly localized spike. Most quantum states are somewhere in between: wave packets that are neither perfectly localized nor perfectly wavelike, having some uncertainty in both position and momentum.

Mathematically, wave packets are formed by superimposing many waves with different wavelengths. A narrow wave packet, localized in a small region, requires superimposing many different wavelengths, giving high momentum uncertainty. A broad wave packet, spread over a large region, can be formed from a narrow range of wavelengths, giving low momentum uncertainty. You cannot make a wave packet both narrow and composed of a single wavelength. The mathematics of Fourier analysis makes this trade off inevitable.

The uncertainty principle applies to other pairs of properties as well. Energy and time form another complementary pair. A state with definite energy must be a time independent wave that has existed forever and will exist forever, giving maximal time uncertainty. A state sharply localized in time is a pulse with very brief duration, which requires superimposing many different energy waves, giving high energy uncertainty. This explains why short lived quantum states have uncertain energies and why spectral lines from atoms have natural widths even in the absence of any measurement imperfections.

Angular momentum components also obey uncertainty relations. If you precisely measure the angular momentum of a quantum system about one axis, the angular momentum components about perpendicular axes become uncertain. This explains many otherwise mysterious features of atomic structure and molecular behavior.

What makes the uncertainty principle so profound is that it reveals fundamental limits on what can be known about physical systems. In classical physics, there are no such limits in principle. Classical limitations on measurement precision arise from practical considerations like instrument quality, thermal noise, or experimental design flaws. With sufficiently advanced technology, classical physics suggests you could measure anything to arbitrary precision. Quantum mechanics says no. Even with perfect instruments, perfect isolation, and perfect measurement techniques, complementary properties cannot simultaneously have definite values.

This uncertainty is not ignorance. It is not that particles have definite positions and momenta but we cannot measure both. The particle genuinely lacks definite values for both properties simultaneously. An electron in an atom does not have a definite position and velocity that we merely cannot determine. It exists in a quantum state where position and momentum are both inherently fuzzy, described by probability distributions rather than definite values.

Consider an electron in the ground state of a hydrogen atom. This state has definite energy but no definite angular position around the nucleus. The electron is not orbiting the nucleus in a classical trajectory like a planet orbits the sun. Instead, the electron's wave function spreads spherically around the nucleus in what is called an s orbital. At any instant, you cannot say the electron is at this location moving in that direction. The electron simply exists as a spherically symmetric probability cloud with no definite position or momentum.

If you measure the electron's position, you will find it at some specific location, but the measurement outcome is random with probabilities determined by the wave function. Before measurement, the electron had no position. After measurement, it has a definite measured position, but now its momentum is completely uncertain and its wave function has changed to reflect the new state of definite position.

The uncertainty principle has enormous practical consequences for atomic structure and stability. Classical physics predicted that atoms should be unstable. An electron orbiting a nucleus is constantly accelerating due to the centripetal force keeping it in orbit. Accelerating charges emit electromagnetic radiation according to Maxwell's equations. The orbiting electron should radiate away its energy and spiral into the nucleus in a fraction of a second. Yet atoms are stable and electrons do not spiral inward.

Quantum mechanics resolves this classical crisis through the uncertainty principle. If an electron were to fall into the nucleus, it would be localized to a very small region defined by the nuclear size, perhaps 10 to the negative 15 meters. Such extreme localization would create enormous momentum uncertainty. The electron would have huge kinetic energy from this momentum uncertainty, much larger than the electrical potential energy attracting it to the nucleus. So the electron cannot fall into the nucleus without gaining so much kinetic energy that it would immediately escape. The lowest energy configuration is a compromise where the electron occupies a relatively small but not infinitesimally small region, giving it the minimum total energy consistent with quantum uncertainty.

This explains why atoms have discrete energy levels rather than a continuum. Only certain wave patterns can fit within the atomic potential while satisfying the Schrödinger equation. These allowed patterns correspond to discrete energy levels. The electron cannot gradually lose energy and spiral inward because there are no allowed quantum states with energies between the discrete levels. The electron can only jump instantaneously from one level to another by emitting or absorbing a photon with energy exactly matching the energy difference.

The uncertainty principle also explains why matter has volume and cannot be compressed indefinitely. When you squeeze matter into a smaller volume, you localize particles to smaller regions, increasing their momentum uncertainty and thus their kinetic energy. This quantum pressure prevents collapse. It explains why even cold matter resists compression. White dwarf stars are supported against gravitational collapse by electron quantum pressure. Neutron stars are supported by neutron quantum pressure. Without quantum uncertainty, all matter would collapse into infinitesimally small points.

Heisenberg originally formulated the uncertainty principle through thought experiments involving measurement. He imagined trying to measure an electron's position by shining light on it and capturing the reflected photon with a microscope. To measure position precisely, you need short wavelength light to resolve small distances. But short wavelength means high energy photons, which impart large momentum kicks to the electron, disturbing its momentum significantly. If you use longer wavelength photons to minimize momentum disturbance, the position measurement becomes imprecise due to diffraction limits.

This measurement interpretation is pedagogically useful but potentially misleading. It suggests uncertainty arises from measurement disturbance, implying particles have definite values that measurement merely fails to reveal. The deeper truth is that uncertainty exists independently of measurement. Even without any measurement, quantum states cannot have definite values for complementary observables simultaneously. The wave function describing a quantum state simply cannot be both a pure position state and a pure momentum state.

The mathematical structure underlying uncertainty is commutation relations. In quantum mechanics, observable are represented by operators acting on wave functions. Position and momentum are represented by operators that do not commute, meaning applying them in different orders gives different results. The commutator of position and momentum equals Planck's constant divided by 2 pi times the imaginary unit. This non zero commutator is the mathematical origin of the uncertainty principle.

Observable with commuting operators can simultaneously have definite values. Observable with non commuting operators cannot. Position and momentum do not commute, so they obey an uncertainty relation. Energy and position do commute in many systems, so a state can have both definite energy and definite position, as occurs for a particle in a box at a specific energy level.

Some interpretations of quantum mechanics modify or reinterpret the uncertainty principle. The pilot wave interpretation, also called de Broglie Bohm theory, postulates that particles do have definite positions and momenta at all times but are guided by a quantum potential derived from the wave function. In this view, uncertainty arises because the particle's initial position is unknown and cannot be controlled, leading to unpredictable trajectories. However, this interpretation requires accepting non local hidden variables and instantaneous action at a distance, features many physicists find unappealing.

The many worlds interpretation sidesteps uncertainty by claiming all possible measurement outcomes occur in different branches of reality. When you measure position, the universe splits into multiple branches, with each branch containing one possible measurement outcome. From your perspective within one branch, the outcome seems random, but no randomness actually exists at the level of the universal wave function. Every possibility happens, just in different branches. This interpretation preserves determinism at the cost of accepting an enormous multiplicity of unobservable parallel universes.

Most physicists simply accept the uncertainty principle as a fundamental feature of nature without committing to a specific interpretation. The mathematical formalism works perfectly regardless of interpretational stance. You can calculate uncertainty products for any quantum state, predict measurement statistics, and design experiments around uncertainty constraints, all without resolving the philosophical questions about what uncertainty really means.

The uncertainty principle has inspired technological innovations precisely because it represents a fundamental limit. Researchers have designed measurement schemes that approach the uncertainty bound as closely as possible, achieving quantum limited precision in applications from gravitational wave detection to atomic clocks. These quantum limited measurements cannot be improved without violating fundamental quantum principles, so reaching them represents a kind of ultimate achievement in precision metrology.

Quantum key distribution for secure communication exploits the uncertainty principle. Any eavesdropper trying to intercept quantum encrypted messages must measure the quantum states being transmitted. This measurement necessarily disturbs the states due to complementarity and uncertainty. The legitimate users can detect this disturbance, revealing the presence of eavesdropping. The security comes from quantum mechanics itself, not from computational complexity or key secrecy. No technological advance can defeat quantum encryption without violating the uncertainty principle, which means defeating quantum mechanics itself.

The uncertainty principle also sets fundamental limits on computing and information processing. The Margolus Levitin theorem states that the minimum time for a quantum system to evolve to an orthogonal state is inversely proportional to its energy, a limit arising from the time energy uncertainty relation. This sets a maximum speed for quantum computation that cannot be exceeded regardless of technological sophistication. Similarly, the Bekenstein bound limits how much information can be stored in a physical system of given energy and size, a limit with deep connections to thermodynamics and black hole physics.

Philosophically, the uncertainty principle represents one of the most radical breaks with classical physics. Classical determinism assumed that the universe evolves according to definite laws, with every event determined by prior events. If you knew the positions and velocities of all particles in the universe at one instant, you could in principle calculate the entire future and past. This clockwork universe had no place for genuine randomness or fundamental indeterminacy.

Quantum uncertainty destroys this deterministic picture. You cannot know both position and momentum precisely, so you cannot specify the exact state of even a single particle, let alone the entire universe. Quantum evolution is deterministic in the sense that Schrödinger's equation determines how wave functions evolve, but measurement outcomes are fundamentally random. The universe is not a clockwork mechanism following pre determined paths but a quantum system evolving according to probabilistic laws, with irreducible uncertainty about outcomes.

This bothered many physicists including Einstein, who spent years searching for ways to avoid or minimize quantum uncertainty. He devised clever thought experiments trying to violate the uncertainty principle or show it was merely apparent rather than fundamental. Each time, deeper analysis revealed subtle effects preserving the uncertainty bound. The EPR paradox, Einstein's most famous attempt to show quantum mechanics was incomplete, ultimately led to Bell's theorem and experimental proof that quantum mechanics really is as strange as it seems, including its fundamental uncertainties.

Modern precision experiments have confirmed the uncertainty principle to extraordinary accuracy. Researchers have prepared quantum states approaching minimum uncertainty, called squeezed states, where one observable is measured with precision below the standard quantum limit at the cost of greater uncertainty in the complementary observable. These squeezed states find applications in gravitational wave detection, where squeezing light reduces noise in the measurement. The uncertainty principle is not violated but exploited, reducing uncertainty in the measured quadrature by accepting increased uncertainty in the unmeasured quadrature.

For students and researchers working with quantum mechanics, the uncertainty principle provides practical guidance. It tells you which measurements can be performed simultaneously and which cannot. It explains why certain quantum states behave in particular ways. It sets benchmarks for measurement precision and information processing speed. It connects abstract mathematical structures like commutators to measurable physical quantities.

For everyone else, the uncertainty principle reveals that reality at its most fundamental level is intrinsically fuzzy. The universe does not have perfect definition like a computer simulation where every variable has an exact value. Quantum reality is probabilistic, with properties that remain genuinely indefinite until measured, complementary observables that cannot simultaneously have definite values, and limits on knowledge that no improvement in technology can ever overcome.

This uncertainty is not a bug in the laws of physics but a feature, and arguably a necessary one. The zero point energy arising from uncertainty prevents atoms from collapsing. It ensures that even at absolute zero temperature, quantum systems retain residual motion, preventing violation of thermodynamic principles. It makes possible many quantum phenomena including tunneling, where particles pass through energy barriers they could never surmount classically. It enables quantum fluctuations that played a crucial role in structure formation in the early universe.

Werner Heisenberg's uncertainty principle stands as one of the greatest intellectual achievements of the twentieth century. In a few simple mathematical inequalities, it captures a profound truth about the nature of reality. Quantum uncertainty is not measurement error, not ignorance, not technological limitation. It is woven into the fabric of quantum mechanics and by extension into the fabric of physical reality itself.

4. Entanglement and Non locality - Einstein's "Spooky Action" Is Real

In 1935, Albert Einstein, along with Boris Podolsky and Nathan Rosen, published a paper that was meant to demonstrate the incompleteness of quantum mechanics. They described a thought experiment, now known as the EPR paradox, involving two particles that interact and then separate to vast distances. According to quantum mechanics, measuring a property of one particle instantaneously determines the corresponding property of the other particle, regardless of the distance separating them. Einstein found this absurd, dismissing it as "spooky action at a distance" that violated the principle of locality, the idea that objects are only directly influenced by their immediate surroundings.

Einstein believed this apparent non locality proved that quantum mechanics was incomplete, that there must be hidden variables determining particle properties all along, with quantum mechanics merely reflecting our ignorance of these hidden details. He was convinced that no reasonable physical theory could allow instantaneous influences across space.

Einstein was wrong. Quantum entanglement, the phenomenon at the heart of the EPR paradox, is completely real and has been demonstrated experimentally thousands of times with ever increasing precision and sophistication. The universe really does exhibit nonlocal correlations that cannot be explained by any local hidden variable theory. This is not speculation or interpretation. It is established experimental fact, proven by tests of Bell's inequality that rule out all possible local realistic explanations of quantum correlations.

Entanglement begins when two or more quantum particles interact in such a way that their quantum states become correlated. After the interaction, even when the particles separate to arbitrarily large distances, they remain described by a single joint quantum state rather than individual independent states. The particles become entangled, sharing a quantum connection that persists regardless of spatial separation.

Consider the simplest case: two electrons prepared in a spin singlet state. Electron spin is a quantum property somewhat analogous to angular momentum, with each electron able to be measured as either spin up or spin down along any chosen axis. In the singlet state, the two electrons are entangled such that their spins always point in opposite directions, but neither electron has a definite spin direction before measurement.

If you measure the first electron and find it spin up along some axis, you instantly know the second electron will be spin down along that same axis, even if the second electron is now on the opposite side of the galaxy. The measurement outcome for the first electron is random, with 50 percent probability for either result. But once you obtain a result, the outcome for the second electron becomes certain, determined by the correlation encoded in the entangled state.

This might not seem so mysterious at first. Perhaps the electrons had definite but unknown spin directions all along, like balls painted red and blue placed in separate boxes. When you open one box and find red, you immediately know the other box contains blue, but nothing spooky happened. The colors were determined when the balls were boxed, and opening one box merely revealed pre existing information.

John Bell proved in 1964 that this local hidden variable explanation cannot reproduce all quantum predictions. Bell derived inequalities that any local realistic theory must satisfy. A local realistic theory is one where particles have definite properties whether or not measured, and where influences cannot propagate faster than light. Bell showed that quantum mechanics predicts violations of these inequalities in certain experimental configurations.

The key is to measure spins along different axes for the two electrons and compare statistics over many repeated trials. If the electrons had definite spin directions determined at the moment of entanglement, correlations between measurements at different angles must satisfy Bell inequalities. Quantum mechanics predicts stronger correlations than any local hidden variable theory can produce, correlations that violate Bell inequalities.

Starting in the 1970s, physicists performed increasingly sophisticated experiments testing Bell inequalities using entangled photons rather than electrons. Photon polarization serves as the measured property, analogous to electron spin. Experimenters prepared entangled photon pairs, sent them to distant measurement stations, measured their polarization along various axes, and compared the statistical correlations.

The results consistently agreed with quantum predictions and violated Bell inequalities, ruling out local hidden variable theories. The photons did not have definite polarization before measurement. The measurement outcomes were genuinely random. Yet the outcomes exhibited correlations stronger than anything possible if each photon had independent pre existing properties.

One might still try to save locality by proposing that the measurement settings themselves are somehow predetermined or that information about measurement choices reaches the other particle through some unknown faster than light signal. Experimenters closed these loopholes through clever experimental designs. They used random number generators to choose measurement settings at the last possible moment, too late for any light speed signal to reach the other measurement station before its measurement was complete. They separated the measurement stations by kilometers and later even hundreds of kilometers, ensuring that measurements occurred in spacelike separated regions where no causal influence could connect them according to relativity.

Every loophole closure confirmed quantum predictions and violated Bell inequalities. In 2015, multiple research groups performed loophole free Bell tests simultaneously closing all major loopholes in single experiments. The conclusion is inescapable: quantum entanglement is real, and it produces correlations that cannot be explained by any local realistic theory.

What exactly happens when you measure an entangled particle? According to standard quantum mechanics, before measurement the entangled pair exists in a superposition of correlated states. Neither particle has a definite spin or polarization. When you measure the first particle, its wave function collapses to a definite outcome, and simultaneously the wave function of the second particle collapses to the correlated outcome. This collapse happens instantaneously regardless of the distance between particles.

Does this violate special relativity, which forbids faster than light signaling? No, because you cannot use entanglement to send messages. The measurement outcome for the first particle is random. You cannot control whether you get spin up or spin down. The second particle's outcome is correlated with yours, but someone measuring the second particle alone sees only random results with no pattern. Only when you later compare notes, through classical communication traveling at or below light speed, do you discover the correlations. So no usable information travels faster than light, and relativity is not violated.

Nevertheless, entanglement represents a kind of non locality, a holism where separated systems cannot be described as having independent properties. The entangled pair must be treated as a single quantum system described by a joint wave function, not as two separate systems that happen to be correlated. This quantum holism conflicts with the classical intuition that spatially separated objects are independent entities.

Einstein hoped that quantum mechanics was merely an effective theory, approximately correct but incomplete, to be replaced eventually by a deeper theory restoring locality and realism. Bell's theorem and subsequent experiments proved this hope was unfounded. Any theory reproducing quantum predictions must either abandon locality, allowing instantaneous influences between separated systems, or abandon realism, denying that particles have definite properties before measurement. Most physicists accept that quantum mechanics abandons realism rather than locality, since it preserves relativistic causality even while exhibiting non local correlations.

Entanglement is not limited to pairs of particles. Physicists have created entangled states involving three, four, dozens, even thousands of particles. As the number of entangled particles grows, the quantum correlations become increasingly complex and the deviation from classical behavior becomes more dramatic. Some proposed quantum computers will rely on entangling hundreds or thousands of quantum bits, or qubits, to perform calculations impossible for classical computers.

Entanglement also need not involve just one property. Particles can be entangled in multiple degrees of freedom simultaneously. Two photons might be entangled in both polarization and which path they took through an interferometer, creating hyper entanglement with richer correlation structure. Particles can be entangled continuously in position and momentum, creating EPR states with perfect correlations in position differences and momentum sums.

The phenomenon of quantum teleportation exploits entanglement to transfer quantum states between distant locations. Suppose Alice wants to send Bob the quantum state of a particle she possesses. She cannot simply measure the particle and tell Bob the results, because measurement destroys the quantum state and provides only limited classical information. Instead, Alice and Bob share an entangled pair of particles prepared in advance. Alice performs a joint measurement on her original particle and her half of the entangled pair. This measurement yields some classical result, which she sends to Bob through ordinary communication. Bob then performs an operation on his half of the entangled pair, conditioned on Alice's measurement result. After this operation, Bob's particle is now in the same quantum state that Alice's original particle had, while Alice's particle is destroyed.

No quantum information travels faster than light in teleportation. The quantum state does not move from Alice to Bob until Alice sends her classical measurement result, which travels at light speed. What does travel faster than light, in some sense, is the correlation: Alice's measurement instantaneously affects Bob's particle, but Bob cannot access that change until he receives Alice's message. Teleportation demonstrates the nonlocal quantum correlations of entanglement while respecting relativistic causality.

Quantum teleportation has been demonstrated experimentally many times, with photons, atoms, and other quantum systems. Researchers have achieved teleportation over increasing distances, first across laboratory tables, then between buildings, then across cities, eventually even from ground stations to satellites in orbit hundreds of kilometers above Earth. These experiments confirm that entanglement persists over macroscopic distances and can be used for practical quantum communication protocols.

Entanglement may play a role in biological systems. Some researchers have proposed that quantum effects including entanglement contribute to photosynthesis, allowing plants to achieve nearly perfect efficiency in transferring energy from light capturing molecules to reaction centers. Bird navigation might exploit quantum entanglement in specialized proteins in the retina, sensitive to Earth's magnetic field through radical pair mechanisms where entangled electron spins respond to magnetic fields. These proposals remain controversial, as biological systems are warm and noisy environments where quantum coherence typically decoheres rapidly. But evidence continues accumulating that some biological processes harness quantum phenomena in sophisticated ways.

Entanglement also appears central to understanding the thermodynamics and information theory of black holes. When a black hole forms and then evaporates through Hawking radiation, the emitted radiation is initially entangled with the black hole's interior. As the black hole shrinks, it carries away information about what fell in, but this information becomes maximally entangled with the outgoing radiation. The question of what happens to this entangled information at the moment the black hole completely evaporates, the black hole information paradox, remains one of the deepest puzzles in theoretical physics, intimately connected to the nature of quantum entanglement in curved spacetime.

Some interpretations of quantum mechanics explain entanglement differently. In the many worlds interpretation, measurement does not collapse wave functions. Instead, the universe splits into multiple branches, with each possible measurement outcome occurring in a different branch. Entangled particles in different branches exhibit correlations because branches remain coherent elements of a larger universal wave function. From within any single branch, correlations appear nonlocal, but the underlying wave function evolution is perfectly local and deterministic at the universal level.

The pilot wave or de Broglie Bohm interpretation maintains that particles always have definite positions guided by a quantum potential derived from the wave function. Entangled particles have positions determined by a nonlocal potential that depends instantaneously on the positions of all particles in the entangled system. This interpretation explicitly embraces nonlocality, with faster than light influences between particle positions, but arranges those influences such that no controllable signals can be sent.

Regardless of interpretational stance, the experimental facts about entanglement are interpretation independent. Entangled systems exhibit correlations that violate Bell inequalities. These correlations cannot be explained by local hidden variables. Quantum mechanics correctly predicts these correlations. Entanglement is real and has been harnessed in numerous applications from quantum cryptography to quantum teleportation to quantum enhanced sensing and metrology.

For a century, physicists have struggled to reconcile entanglement with our intuitions about space, separation, and causation. We naturally think of spatially separated objects as independent, with properties they possess regardless of what happens elsewhere. Quantum mechanics says this classical picture is fundamentally wrong. Entangled systems must be treated holistically, as single quantum objects described by joint wave functions that encode correlations transcending spatial separation.

This does not mean that space is an illusion or that everything in the universe is directly connected to everything else. Most quantum systems are not entangled with each other. Decoherence rapidly destroys entanglement between systems interacting with noisy environments. To observe entanglement, you must carefully isolate quantum systems and perform measurements before environmental interactions destroy quantum coherence. But the possibility of entanglement, even if difficult to maintain in practice, reveals something profound about the quantum structure of reality.

Einstein called entanglement "spooky action at a distance" because he found it incompatible with his deepest physical intuitions. He was right that it is spooky, in the sense of being deeply counterintuitive and unlike anything in classical physics. He was wrong that it indicated incompleteness in quantum mechanics. Experiments have proven beyond any reasonable doubt that quantum entanglement is a real feature of nature, that the universe really does permit correlations stronger than anything possible in a local realistic world, and that quantum mechanics correctly describes these correlations.

The discovery and experimental confirmation of entanglement ranks among the most profound findings in the history of physics. It demonstrates that reality is fundamentally nonlocal in a precisely defined sense, that quantum systems cannot be understood by analyzing their parts in isolation, and that the universe exhibits a subtle form of holism at odds with classical separability. Entanglement is not just a curiosity but a resource, enabling quantum technologies from secure communication to computational speedups to enhanced precision measurements.

Living in a quantum universe means accepting that separated systems can be correlated in ways that defy classical explanation, that measuring one particle can instantaneously determine properties of another particle across the cosmos, and that this apparent action at a distance is not magic or error but simply how quantum reality operates. Einstein spent the last decades of his life trying to prove quantum mechanics wrong about this. Experiments proved him wrong instead. Quantum entanglement is real, and it is here to stay.

II

II. Relativity and the Cosmos

Having explored the quantum realm where reality becomes probabilistic, uncertain, and non local, we now turn to the physics of the very fast and the very massive. This is the domain of Einstein's relativity, where space and time lose their absolute character, where gravity reveals itself as the curvature of spacetime itself, and where extreme conditions create some of the most bizarre objects in the universe: black holes.

5. Time Is Not Universal - Why Astronauts Age Slower Than Earth

Isaac Newton opened his masterwork Principia Mathematica with definitions of absolute space and absolute time. Absolute space, according to Newton, remains always similar and immovable, providing a fixed reference frame for all motion. Absolute time flows equably forward, independent of anything external, ticking at the same rate everywhere in the universe for all observers. These concepts seemed self evident, matching everyday experience where time flows steadily and space provides a stable stage for physical events.

For more than two centuries, Newton's framework prevailed. Classical mechanics built on absolute space and time achieved stunning successes, explaining planetary orbits, predicting the existence of Neptune from perturbations in Uranus's orbit, describing the motion of projectiles, pendulums, and spinning tops. The edifice of physics seemed nearly complete by the late 1800s.

Then cracks appeared. The most serious involved light and electromagnetic waves. James Clerk Maxwell had unified electricity, magnetism, and light into a single theoretical framework, Maxwell's equations, which predicted that electromagnetic waves propagate at a specific speed, approximately 300,000 kilometers per second. But speed relative to what? In Newtonian physics, velocities add. If you throw a ball forward at 20 kilometers per hour from a train moving 100 kilometers per hour, observers on the ground measure the ball's speed as 120 kilometers per hour. Similarly, light emitted forward from a moving source should travel faster relative to stationary observers than light emitted from a stationary source.

Experiments showed otherwise. The most famous was the Michelson Morley experiment of 1887, which attempted to detect Earth's motion through the hypothetical luminiferous ether, the medium thought to carry light waves. The experimenters measured light speed in different directions as Earth orbited the Sun, expecting to find variations as Earth sometimes moved toward the light source and sometimes away. They found no variation whatsoever. Light traveled at exactly the same speed regardless of the direction of measurement or Earth's motion.

This null result baffled physicists for nearly two decades. Some proposed that Earth dragged the ether along with it. Others suggested that objects moving through the ether physically contract in the direction of motion by exactly the amount needed to hide any velocity dependence of light speed. These explanations felt contrived, special fixes to preserve absolute space and time.

Albert Einstein took a radically different approach in 1905. Rather than trying to explain away the constancy of light speed, he embraced it as a fundamental principle. He proposed that the speed of light in vacuum is the same for all observers, regardless of their motion or the motion of the light source. Combined with the principle that the laws of physics are identical for all observers moving at constant velocities, this postulate led inexorably to special relativity and the overthrow of absolute space and time.

The key insight is that observers moving relative to each other cannot agree on whether distant events are simultaneous. Imagine a train moving past a station platform. Lightning strikes the front and back of the train simultaneously according to platform observers, who see the light from both strikes reach the train's midpoint at the same moment. But observers on the train see the light from the front strike arrive before light from the rear strike, because the train is moving toward the front strike and away from the rear strike, and light travels at the same speed for all observers. Since light speed is the same for both platform and train observers, they cannot agree on whether the strikes were simultaneous.

This relativity of simultaneity has profound consequences. If observers cannot agree on whether events are simultaneous, they cannot agree on time intervals either. A clock moving relative to you appears to tick slower than your own identical clock, an effect called time dilation. A meter stick moving relative to you appears shorter in the direction of motion than an identical meter stick at rest, an effect called length contraction. These are not optical illusions or measurement errors. They are real physical effects arising from the structure of spacetime.

The mathematics of special relativity involves the Lorentz transformation, equations relating the space and time coordinates measured by observers in relative motion. The transformation depends on the relative velocity and the speed of light. For velocities much smaller than light speed, the transformation reduces to the classical Galilean transformation, explaining why everyday experience misleadingly suggests absolute space and time. But as velocities approach light speed, relativistic effects become dramatic.

Time dilation has been confirmed experimentally countless times. Muons are subatomic particles created when cosmic rays strike the upper atmosphere. They decay with a half life of about 2 microseconds when at rest. At the speeds reached by cosmic ray muons, about 98 percent of light speed, they should decay long before reaching Earth's surface even at light speed. Yet muons are detected at sea level in abundance. From Earth's reference frame, the muons' internal clocks run slow due to time dilation, giving them enough time to reach the ground before decaying. From the muons' reference frame, they live their usual 2 microseconds, but Earth rushes toward them at high speed and the atmosphere is length contracted to a thin layer, allowing them to traverse it before decaying.

Both perspectives are equally valid. There is no absolute answer to how long the muon lived. Measured by Earth clocks, it lived much longer than 2 microseconds. Measured by its own internal clock, it lived its natural 2 microseconds. Both statements are true in their respective reference frames. Time really is relative.

Particle accelerators routinely accelerate particles to speeds exceeding 99.9 percent of light speed, creating dramatic time dilation factors. At CERN's Large Hadron Collider, protons circle the 27 kilometer ring at 99.9999991 percent of light speed. At this velocity, time dilation is so extreme that from the protons' perspective, their journey around the ring takes only a tiny fraction of the time measured by laboratory clocks. Accelerator designers must account for this time dilation to correctly calculate particle trajectories and collision rates.

Human astronauts experience much more modest time dilation, but it is still measurable. Astronauts aboard the International Space Station orbit Earth at about 7.7 kilometers per second. At this speed, time passes about 0.007 seconds slower per year in the astronauts' reference frame compared to Earth's surface. Over a six month mission, an astronaut ages about 0.0035 seconds less than people remaining on Earth. This tiny effect has been confirmed by precise atomic clocks flown on aircraft and satellites, which gain or lose time relative to ground clocks exactly as relativity predicts.

The Global Positioning System depends critically on relativistic corrections. GPS satellites orbit at about 20,000 kilometers altitude, moving at roughly 3.9 kilometers per second. At this speed, special relativistic time dilation causes satellite clocks to run slow by about 7 microseconds per day. However, general relativistic effects from Earth's gravitational field, which we will discuss in the next chapter, cause satellite clocks at higher altitude to run fast by about 45 microseconds per day. The net effect is that satellite clocks gain about 38 microseconds per day relative to ground clocks. Since GPS determines position by timing how long radio signals take to travel from satellites to receivers, and since radio signals travel at light speed, a timing error of 38 microseconds translates to a position error of about 11 kilometers per day. Without relativistic corrections, GPS would be completely useless within hours.

Time dilation also implies that traveling close to light speed allows time travel to the future, at least in a sense. Suppose astronauts accelerate to 99.99 percent of light speed and travel to a star 10 light years away, then return. From Earth's perspective, the journey takes about 20 years. But due to time dilation at that speed, only about 2.8 years pass on the astronauts' clocks. When they return, they have aged 2.8 years while Earth has aged 20 years. The astronauts have effectively traveled 17.2 years into Earth's future.

This is not reversible. You can travel to the future by moving fast, but you cannot return to your starting time. Time dilation is not time travel in the science fiction sense of jumping between arbitrary times. It is simply that different observers measure different amounts of elapsed time between events, and fast moving observers measure less elapsed time than stationary observers.

The theoretical maximum for this effect occurs as velocity approaches light speed. At exactly light speed, time dilation becomes infinite. If you could travel at light speed, which is impossible for massive objects, no time would pass for you at all. From your perspective, you would traverse any distance instantaneously. A photon emitted from a distant galaxy and absorbed by your eye 13 billion years later experiences zero elapsed time from its own perspective. This infinite time dilation at light speed is one reason why nothing with mass can reach light speed. Accelerating to light speed would require infinite energy and result in time stopping completely in your reference frame.

Length contraction is the spatial counterpart to time dilation. A meter stick moving past you at high speed appears shorter than one meter in the direction of motion. At 87 percent of light speed, it appears half a meter long. At 99.5 percent of light speed, it appears one tenth of a meter long. Like time dilation, length contraction is completely symmetric. From the meter stick's perspective, it has normal length but you and your measuring apparatus are length contracted.

This might seem paradoxical. How can each observer measure the other as contracted? The resolution is that length measurements require simultaneous measurement of both ends of the object, and simultaneity is relative. When you measure both ends simultaneously in your reference frame, those measurements are not simultaneous in the meter stick's frame. When the meter stick's frame measures both ends simultaneously, those measurements are not simultaneous in your frame. Each observer correctly measures the other as length contracted using their own notion of simultaneity.

These effects combine in the relativity of simultaneity, time dilation, and length contraction are different aspects of the same phenomenon: the unification of space and time into spacetime. Spacetime is a four dimensional structure where time is treated as a fourth dimension alongside the three spatial dimensions. Events are specified by four coordinates: three spatial coordinates and one time coordinate. The interval between events, combining spatial and temporal separation, is the fundamental invariant quantity that all observers agree on.

Mathematically, spacetime has a metric, a rule for calculating intervals between events. In special relativity, flat spacetime away from gravitating masses, the metric is the Minkowski metric, which differs from Euclidean geometry by treating time differently from space. The spacetime interval between two events can be timeline if one event can causally influence the other, space like if they are too far apart for light to travel between them, or light like if light can exactly connect them. All observers agree on whether intervals are timeline, space like, or light like, even though they disagree on individual space and time separations.

Causality is preserved in relativity through the cosmic speed limit. Information and causal influences cannot propagate faster than light. Events with space like separation cannot influence each other. If they could, it would be possible to send signals to your own past, creating paradoxes. Relativity forbids faster than light travel precisely to prevent these paradoxes. The speed of light is not merely a fast speed but represents the maximum rate at which causation can propagate through spacetime.

This cosmic speed limit has enormous implications for interstellar travel and communication. The nearest star to Earth, Proxima Centauri, is 4.24 light years away. Even traveling at light speed, which is impossible for spaceships, would require 4.24 years each way. At achievable speeds for current technology, perhaps 0.01 percent of light speed for spacecraft like Voyager, the journey would take tens of thousands of years. For humans to explore even nearby stars requires either extremely long mission durations with generation ships, suspended animation technologies that do not yet exist, or breakthrough physics allowing faster than light travel, which likely violates fundamental principles.

Some science fiction concepts like warp drives and wormholes attempt to circumvent the light speed limit by warping spacetime itself rather than moving through spacetime. These ideas are not obviously impossible, but they typically require exotic matter with negative energy density, which may not exist, and they often lead to time travel paradoxes if actually constructed. Most physicists are skeptical that faster than light travel is possible without violating causality.

The unification of space and time into spacetime is one of the most profound insights of modern physics. Space and time are not separate, absolute backdrops for physical processes. They are interrelated aspects of a four dimensional structure that behaves differently than our three dimensional spatial intuition suggests. Different observers slice spacetime into space and time differently, depending on their motion. There is no absolute present moment dividing past from future. The division between past, present, and future is observer dependent.

This has philosophical implications that troubled Einstein himself. In a famous letter consoling the widow of his friend Michele Besso, Einstein wrote that the distinction between past, present, and future is only a stubbornly persistent illusion. If all events exist timelessly in the four dimensional spacetime, laid out eternally like landscapes on a map, then perhaps time does not truly flow. Perhaps change is an illusion arising from our limited perspective as beings perceiving reality one moment at a time.

Not all physicists accept this block universe view. Some argue that relativity describes only the structure of spacetime measurements, not the fundamental nature of time. They contend that there is objective change and becoming, even if observers disagree about simultaneity. These debates continue, but what is certain is that special relativity forces us to radically reconceptualize space and time.

Time is not absolute. Time is not universal. Clocks moving relative to each other measure different elapsed times between the same events. This is not error or illusion. It is the actual structure of spacetime revealed by relativity. Astronauts really do age slower than people on Earth. Fast moving particles really do live longer than stationary particles. And you really could travel to Earth's future by accelerating to near light speed and returning, though you could never travel back to your own past.

Newton's absolute time is gone, replaced by Einstein's relativistic time that flows at different rates depending on velocity and, as we will see next, depending on gravitational fields as well. This is the universe we actually inhabit, strange and beautiful and profoundly different from the universe of everyday intuition.

6. $E=mc^2$ and the Power Within Matter - What Einstein's Equation Actually Means

Of all the equations in physics, E equals mc squared is undoubtedly the most famous. It appears on t-shirts, coffee mugs, and posters. It represents Einstein and relativity in popular culture. But what does it actually mean, and why is it so important?

The equation states that energy E equals mass m times the speed of light c squared. Since the speed of light is enormous, about 300 million meters per second, and this quantity is squared, making it even more enormous, the equation implies that even a tiny amount of mass corresponds to a tremendous amount of energy. One kilogram of mass, if completely converted to energy, would release about 90 million trillion joules, equivalent to the energy in 21 megatons of TNT.

This mass energy equivalence is perhaps the most famous consequence of special relativity, though Einstein did not include it in his original 1905 paper on special relativity. He derived it later that year in a brief follow up paper titled "Does the Inertia of a Body Depend on Its Energy Content?" The answer was yes. A body's mass increases when it gains energy and decreases when it loses energy, with the relationship given by E equals mc squared.

To understand where this equation comes from, we must return to the relationship between energy and momentum in relativity. In classical physics, a moving object has kinetic energy equal to one half its mass times velocity squared, and momentum equal to mass times velocity. These are separate quantities, related but distinct.

In special relativity, energy and momentum form a four dimensional energy momentum vector, analogous to how position and time form a four dimensional spacetime vector. The invariant magnitude of this four vector, the quantity all observers agree on regardless of their motion, relates to the rest mass of the object. From this four vector structure, it follows that the total energy of an object equals the square root of the sum of its rest mass energy squared plus its momentum times c squared.

For an object at rest with zero momentum, this reduces to E equals mc squared, where m is the rest mass. This rest energy represents energy the object possesses simply by virtue of having mass, independent of any motion. An object gains additional kinetic energy when it moves, but even at rest it has energy locked up in its mass.

This was revolutionary. In classical physics, mass and energy were separate conserved quantities. Matter could neither be created nor destroyed, energy could neither be created nor destroyed, but mass and energy were fundamentally different. Relativity revealed them as two forms of the same underlying quantity. Mass can be converted to energy, energy can be converted to mass, and the conversion rate is c squared.

The first experimental confirmation came from nuclear physics. Careful measurements showed that atomic nuclei weigh slightly less than the sum of their constituent protons and neutrons. A helium nucleus containing two protons and two neutrons weighs about 0.7 percent less than two free protons plus two free neutrons. The missing mass, called the mass defect, has been converted to binding energy holding the nucleus together. This binding energy can be released if the nucleus is disassembled or calculated using E equals mc squared from the mass defect.

Nuclear fission exploits this mass energy conversion. When a uranium nucleus splits into two lighter nuclei plus neutrons, the products weigh slightly less than the original uranium nucleus. The missing mass has been converted to kinetic energy of the fission fragments and neutrons. This energy release is what powers nuclear reactors and fission bombs. A typical fission reaction releases about 200 million electron volts of energy, corresponding to a mass decrease of about 0.1 percent of the original nucleus.

Nuclear fusion is even more efficient. When four hydrogen nuclei fuse to form one helium nucleus in the core of the Sun, about 0.7 percent of the mass is converted to energy. The Sun converts about 4 million tons of mass to energy every second through fusion reactions, generating the light and heat that sustain life on Earth. Over its 10 billion year lifetime, the Sun will convert about 0.03 percent of its total mass to energy through fusion.

Fusion bombs and experimental fusion reactors attempt to harness this process on Earth. The challenge is that fusion requires bringing positively charged nuclei close enough together to overcome electromagnetic repulsion and allow the strong nuclear force to bind them. In the Sun, this occurs naturally due to immense pressure and temperature in the core. On Earth, achieving similar conditions requires either enormous energy input to heat a confined plasma to hundreds of millions of degrees, as in tokamak fusion reactors, or using a fission bomb to create the necessary temperature and pressure, as in thermonuclear weapons.

Antimatter annihilation provides the most complete conversion of mass to energy. Every particle has a corresponding antiparticle with opposite charge but identical mass. When a particle meets its antiparticle, they annihilate completely, converting 100 percent of their combined mass to energy, usually in the form of photons or other particles. An electron meeting a positron, its antiparticle, produces two photons carrying the combined rest energy of both particles.

Antimatter occurs naturally but rarely. Cosmic rays striking the atmosphere produce antiparticles among their collision products. Certain radioactive decay processes produce positrons. Particle accelerators create antimatter routinely by colliding high energy particles, producing particle antiparticle pairs from the collision energy. But antimatter is difficult to store because it annihilates upon contact with ordinary matter. Specially designed electromagnetic traps can confine antiparticles in vacuum, isolating them from any matter. Even so, no one has successfully stored significant quantities of antimatter for extended periods.

The practical difficulty of producing and storing antimatter makes it useless as an energy source despite science fiction depictions of antimatter engines. Creating antiparticles requires at least as much energy as you get back from annihilation, plus substantial overhead for the accelerators and storage systems. Antimatter serves as an energy storage medium at best, and a very inefficient one. For energy generation, fusion offers much more practical potential.

E equals mc squared also appears in particle physics in reverse: energy can be converted to mass. When high energy particles collide in accelerators, their kinetic energy can materialize as new particles that did not exist before the collision. The more energy available, the more massive the particles that can be created. This is how physicists discovered most fundamental particles. They built higher energy accelerators, collided particles at increasing energies, and observed what new particles emerged.

The Higgs boson, discovered at the Large Hadron Collider in 2012, has a mass of about 125 billion electron volts divided by c squared. To create a Higgs, you need a collision with at least this much energy available in the center of mass frame. The LHC accelerates protons to 6.5 trillion electron volts each, so when two protons collide head on, about 13 trillion electron volts is available, enough to create a Higgs plus other particles. The collision energy materializes as particles according to E equals mc squared, with energy converted to mass.

This mass from energy conversion is also how all matter originally formed. In the first moments after the Big Bang, the universe was filled with tremendously high energy radiation. As the universe expanded and cooled, this energy condensed into particle antiparticle pairs, which then annihilated back to radiation, which condensed again to particles. A slight asymmetry between matter and antimatter meant that after most particles annihilated, a small excess of matter remained. Every proton, neutron, and electron in the universe today is descended from this tiny matter excess in the early universe, all originally created from pure energy through E equals mc squared.

The equation also explains why nothing with mass can reach light speed. As an object accelerates toward light speed, its kinetic energy increases without bound. This energy contributes to the object's mass, making it harder to accelerate further. The closer you get to light speed, the more energy is required for additional acceleration, approaching infinite energy to reach exactly light speed. Only massless particles like photons can travel at light speed, and they must always travel at light speed, never slower.

Photons have zero rest mass but still have energy and momentum. The full relativistic energy momentum relationship applies to photons with rest mass set to zero, giving photon energy equal to momentum times c. Since photons have energy, and E equals mc squared relates energy to mass, shouldn't photons have mass? The resolution is that photons have no rest mass, no mass in their own reference frame, but they do have relativistic mass, a term some physicists use for energy divided by c squared. Modern practice avoids the term relativistic mass and simply says that photons are massless particles carrying energy and momentum.

Black holes provide the most extreme demonstration of mass energy equivalence. When a massive star exhausts its nuclear fuel and collapses, vast amounts of gravitational potential energy are released. Some of this energy escapes as a supernova explosion, but the remaining energy can be so concentrated that it warps spacetime enough to create an event horizon, a boundary from within which nothing can escape. The black hole has a mass, measurable by its gravitational field, but this mass is not localized matter in the classical sense. It is pure gravitational energy so concentrated that it curves spacetime into a closed surface.

Some theoretical black hole formation mechanisms involve no matter at all, only energy. A sufficiently intense laser pulse, if focused to a small enough region, could in principle create so much electromagnetic energy density that it forms a black hole called a kugelblitz. No one has the technology to focus lasers intensely enough for this, but it is allowed in principle by E equals mc squared. Pure energy can gravitate and curve spacetime just like mass does because energy and mass are equivalent.

E equals mc squared also has profound implications for cosmology. The total energy of the universe includes not just mass energy but also gravitational potential energy, kinetic energy, and various field energies. In an expanding universe, energy is not conserved in the usual sense because space itself is expanding and energy can be transferred to or from the gravitational field. The cosmological constant or dark energy, which drives accelerated expansion, can be thought of as vacuum energy, a form of energy inherent in empty space itself, contributing to the universe's total mass energy.

The equation tells us that matter is condensed energy, energy is liberated mass, and the two are interchangeable at the exchange rate of c squared. This unification of mass and energy stands as one of Einstein's greatest insights. It explains nuclear power, shows how the Sun shines, allows particle creation and annihilation, and reveals that the universe is filled with energy in multiple forms, all connected through relativistic physics.

When you look at a rock, you are looking at condensed energy. The atoms in the rock, the electrons and nuclei, represent energy that has taken on the form of matter. If you could convert that rock entirely to energy, it would release an explosion dwarfing any human weapon. That this energy remains locked up in matter, not easily accessible, is fortunate for our survival. But the energy is there, demonstrating that mass and energy are not different substances but different manifestations of the same underlying reality.

E equals mc squared is not just a famous equation. It is a window into the deep connection between mass and energy, a connection that becomes apparent only when you take seriously the implications of special relativity. It shows that the universe is more unified than classical physics suggested, with quantities we thought separate and distinct revealed as aspects of a more fundamental unity. This is the power of relativity: it reveals connections and symmetries hidden from classical perspectives, showing us a universe more elegant and integrated than our everyday experience would suggest.

7. Gravity as Curved Space - How Gravity Is Not a Force But Geometry

Isaac Newton's law of universal gravitation, published in 1687, stated that every mass attracts every other mass with a force proportional to the product of their masses and inversely proportional to the square of the distance between them. This law worked spectacularly well, explaining planetary orbits, the Moon's motion, tides, falling objects, and countless other phenomena. For more than two centuries, Newtonian gravity was considered essentially correct, a crowning achievement of physics.

But Newton himself was troubled by one aspect of his theory: action at a distance. How does the Sun reach across 150 million kilometers of empty space to pull on Earth? How does this gravitational force know instantly if the Sun's mass changes? Newton candidly admitted he could frame no hypothesis about the mechanism. He simply described mathematically how gravity behaves without explaining what gravity is.

Albert Einstein resolved this mystery in 1915 with general relativity, his theory of gravitation. General relativity explains gravity not as a force acting across space but as a consequence of the curvature of spacetime itself. Mass and energy warp the fabric of spacetime, and objects move along the straightest possible paths, called geodesics, through this curved spacetime. What we perceive as gravitational attraction is actually objects following geodesics in warped spacetime.

To understand this, imagine a two dimensional analogy: a stretched rubber sheet representing flat space. Place a bowling ball on the sheet. The ball creates a depression, warping the sheet. Now roll a marble across the sheet near the bowling ball. The marble's path curves around the depression, not because the bowling ball exerts a force on it, but because the marble follows the curvature of the sheet itself.

In this analogy, the rubber sheet is two dimensional space, the bowling ball represents a massive object like the Sun, and the marble represents a smaller object like Earth. The marble orbits the bowling ball not because of a force pulling it but because it is following the natural geometry of the curved sheet. General relativity extends this idea to four dimensional spacetime. Mass and energy curve spacetime, and objects move along geodesics in this curved spacetime.

Einstein arrived at general relativity through a thought experiment he later called the happiest thought of his life. Imagine someone falling freely in a gravitational field, perhaps in an elevator with a cut cable. During the fall, the person feels weightless. Objects released inside the elevator float freely relative to the elevator. From the person's perspective inside the elevator, gravity has vanished. This led Einstein to realize that

I'll continue with Chapter 4 and the remaining chapters. Would you like me to proceed with the complete manuscript, or would you prefer to review what I've written so far first?

This equivalence principle became the foundation of general relativity. In a small region of spacetime, you cannot distinguish between being in a gravitational field and being in an accelerating reference frame. An astronaut in a rocket accelerating at 9.8 meters per second squared feels exactly the same as someone standing on Earth's surface experiencing 9.8 meters per second squared gravitational acceleration. Conversely, someone in free fall experiences weightlessness identical to floating in deep space far from any gravitating masses.

From this principle, Einstein deduced that gravity must curve spacetime. Suppose you shine a horizontal laser beam across an accelerating rocket.

From outside the rocket, the beam travels straight while the rocket accelerates upward, so the beam hits the opposite wall lower than where it started. From inside the rocket, the beam appears to curve downward. By the equivalence principle, light must also curve downward in a gravitational field. But light always follows the straightest possible path through spacetime, a null geodesic. If light curves in a gravitational field, then spacetime itself must be curved.

Developing the mathematics to describe curved spacetime took Einstein nearly a decade. He needed to learn differential geometry, the mathematics of curved spaces and manifolds, largely with help from his mathematician friend Marcel Grossmann. The result was the Einstein field equations, a set of ten coupled nonlinear partial differential equations relating the curvature of spacetime to the distribution of mass and energy within it.

The field equations are extraordinarily complex and have been solved exactly only in special cases with high symmetry. The simplest exact solution, found by Karl Schwarzschild in 1916, describes the spacetime geometry around a spherically symmetric non rotating mass. This Schwarzschild solution describes the space around stars, planets, and eventually black holes. More complex solutions describe rotating masses, gravitational waves, expanding universes, and other configurations.

The Einstein field equations encapsulate a profound idea: matter tells spacetime how to curve, and spacetime tells matter how to move. The distribution of mass and energy determines the curvature of spacetime through the field equations. Objects then move along geodesics, the straightest possible paths, through this curved spacetime. There is no gravitational force pulling objects together. Objects simply follow the natural geometry of spacetime, and that geometry is warped by mass and energy.

This explains why all objects fall at the same rate in a gravitational field regardless of their composition. In Newtonian physics, this equality of gravitational and inertial mass was a mysterious coincidence. Why should the mass determining how much gravitational force an object feels be exactly proportional to the mass determining how hard it is to accelerate? In general relativity, there is no coincidence. All objects follow geodesics in curved spacetime, and geodesics depend only on the geometry, not on the object's properties. A feather and a hammer fall at the same rate in vacuum because both follow the same geodesic, both traveling along the same path through curved spacetime.

General relativity makes several predictions that differ from Newtonian gravity, and all have been confirmed experimentally. The first involves the orbit of Mercury. Astronomers had long observed that Mercury's orbit precesses, meaning its elliptical path slowly rotates in space. Most of this precession was explainable by gravitational perturbations from other planets, but about 43 arcseconds per century remained unexplained. General relativity predicted exactly this excess precession as a consequence of the curvature of spacetime near the Sun. When Einstein calculated the precession and found agreement with observations, he later said his heart palpitated and he felt something actually snap inside him.

The second prediction was gravitational light deflection. Light passing near the Sun should curve due to spacetime curvature. Einstein calculated that starlight grazing the Sun's surface should deflect by about 1.75 arcseconds. This prediction was tested during the solar eclipse of 1919 by Arthur Eddington, who photographed stars near the Sun during totality when the Sun's glare was blocked. The stars appeared displaced from their normal positions by amounts matching Einstein's prediction. This dramatic confirmation made Einstein internationally famous virtually overnight.

The third prediction was gravitational redshift. Light climbing out of a gravitational field loses energy, causing its wavelength to increase and frequency to decrease. This is not because photons slow down, as they always travel at c, but because time runs slower deeper in gravitational fields. A photon emitted from the surface of a massive object like a white dwarf star is redshifted when observed from far away. This effect has been measured in laboratory experiments on Earth and observed in light from white dwarfs and neutron stars.

Gravitational time dilation, the slowing of time in gravitational fields, is one of general relativity's most striking predictions. Clocks deeper in gravitational wells run slower than clocks higher up. On Earth's surface, time runs slightly slower than at high altitude. The effect is tiny, only about 1 microsecond per century difference between sea level and mountain tops, but it is measurable with atomic clocks and must be accounted for in GPS satellites.

This gravitational time dilation has been confirmed in numerous experiments. In the Pound Rebka experiment of 1959, gamma ray photons were sent up and down a 22.5 meter tower at Harvard. The photons traveling upward were redshifted, while those traveling downward were blueshifted, by amounts precisely matching general relativity's predictions. In 1976, scientists launched a rocket carrying an atomic clock to an altitude of 10,000 kilometers. The clock ran faster than ground clocks exactly as predicted by general relativity.

Black holes represent the most extreme consequence of spacetime curvature. If enough mass is compressed into a small enough region, spacetime curves so severely that it closes in on itself, creating a region from which nothing can escape. The boundary of this region is the event horizon. Outside the event horizon, escape is possible though difficult. Inside the event horizon, all paths through spacetime lead inward to the center regardless of how you accelerate or what direction you try to go. Even light cannot escape because even lightlike geodesics point inward.

The curvature at a black hole's center, the singularity, becomes infinite according to general relativity. This infinite curvature suggests that general relativity itself breaks down there, that a quantum theory of gravity is needed to describe physics at the singularity. But the event horizon shields this breakdown from outside observers, an idea called cosmic censorship. Naked singularities, if they exist, would represent regions where spacetime curvature becomes infinite yet remains visible from far away, potentially causing causality violations. Most physicists believe nature forbids naked singularities, though this remains unproven.

Gravitational waves are another prediction of general relativity. Accelerating masses create ripples in spacetime curvature that propagate outward at the speed of light. These waves stretch and squeeze space as they pass, alternately lengthening and shortening the distance between objects. The effect is extraordinarily small. Even gravitational waves from colliding black holes cause strains in spacetime of only about one part in 10 to the 21st power, far too small to notice without extraordinarily sensitive detectors.

In 2015, the Laser Interferometer Gravitational Wave Observatory, LIGO, detected gravitational waves for the first time. The waves came from two black holes orbiting each other and merging 1.3 billion light years away. As the waves passed through Earth, they changed the length of LIGO's four kilometer long detector arms by less than one thousandth the diameter of a proton. This minuscule signal was detected by measuring interference between laser beams traveling along perpendicular arms. Since then, LIGO and other gravitational wave detectors have observed dozens of black hole mergers and neutron star mergers, opening an entirely new way to observe the universe.

Gravitational lensing occurs when light from distant objects passes through regions of strong spacetime curvature. The curved spacetime acts like a lens, bending light rays and creating multiple images, magnified images, or distorted arcs and rings. Astronomers use gravitational lensing to study distant galaxies, measure the masses of galaxy clusters, and search for dark matter. Some of the most distant objects ever observed are visible only because their light is magnified by gravitational lensing from intervening galaxy clusters.

General relativity also describes the large scale structure and evolution of the universe itself. In 1917, Einstein applied his field equations to the universe as a whole, assuming it is homogeneous and isotropic on large scales. He found that his equations predicted a universe that either expands or contracts, never remaining static. This troubled Einstein, as astronomers at the time believed the universe was eternal and unchanging, so he added a cosmological constant term to his equations to allow a static solution.

In 1929, Edwin Hubble discovered that distant galaxies are receding from us with velocities proportional to their distances, showing that the universe is expanding. Einstein quickly abandoned the cosmological constant, reportedly calling it his greatest blunder. The expanding universe is now understood as space itself expanding, carrying galaxies apart, not galaxies moving through static space. General relativity provides the framework for modern cosmology, describing how the universe evolved from an initial hot dense state through 13.8 billion years of expansion to its current large scale structure.

Interestingly, the cosmological constant has made a comeback. Observations since the 1990s show that the universe's expansion is accelerating, driven by what is called dark energy. The simplest explanation for dark energy is a positive cosmological constant, a form of energy inherent in empty space itself. Einstein's supposed blunder may have been prescient after all, though for the wrong reasons. The universe is not static, but it does appear to have a cosmological constant driving accelerated expansion.

General relativity also allows for exotic solutions including wormholes, shortcuts through spacetime connecting distant regions. The mathematics permits such solutions, but they typically require exotic matter with negative energy density to hold the wormhole open against collapse. Whether such matter exists and whether traversable wormholes are possible remains unknown. Most physicists are skeptical that wormholes suitable for travel exist or can be created.

Time travel to the past is another exotic possibility in general relativity. Certain solutions to the Einstein field equations, such as rotating black holes or infinitely long rotating cylinders, contain closed timelike curves, paths through spacetime that loop back to their starting point in time. These solutions typically require unrealistic conditions like infinite extent or naked singularities, and they often lead to paradoxes. Whether the universe actually permits time travel to the past remains one of the deepest open questions in theoretical physics.

Despite its successes, general relativity is incomplete. It does not incorporate quantum mechanics and breaks down at singularities where curvature becomes infinite. Physicists expect that a quantum theory of gravity will replace general relativity in extreme conditions like black hole singularities and the Big Bang. String theory, loop quantum gravity, and other approaches attempt to develop such a theory, but no complete quantum gravity theory has been experimentally confirmed.

Nevertheless, within its domain of applicability, general relativity is extraordinarily successful. It has passed every experimental test, from subtle corrections to Mercury's orbit to the detection of gravitational waves from colliding black holes. It provides the framework for understanding cosmic structure and evolution. It reveals that gravity is not a mysterious force acting at a distance but a natural consequence of the geometry of spacetime itself.

This geometrization of gravity represents one of the most beautiful ideas in physics. Rather than treating gravity as something separate from spacetime, general relativity shows that gravity is spacetime. Mass and energy determine how spacetime curves, and this curvature determines how objects move. The universe is not objects moving through space and time as a fixed backdrop. The universe is spacetime itself, dynamically curved by its contents, with that curvature determining the motion of everything within it.

When you fall, you are not being pulled by a force. You are following the straightest possible path through curved spacetime. When Earth orbits the Sun, it is not held in orbit by a force reaching across space. It is following a geodesic through spacetime warped by the Sun's mass. Gravity is geometry, and general relativity is the mathematics describing that geometry.

8. Black Holes, Wormholes, and Singularities - Where Physics Meets Infinity

In November 1915, just weeks after Einstein published his general theory of relativity, the German astronomer Karl Schwarzschild found an exact solution to Einstein's field equations. Schwarzschild was serving in the German army on the Russian front during World War I when he performed these calculations. Tragically, he died just months later from a disease contracted in the trenches. But his solution, the Schwarzschild metric, became the foundation for understanding one of the most extreme predictions of general relativity: black holes.

The Schwarzschild solution describes the spacetime geometry around a spherically symmetric, non rotating mass. For ordinary stars and planets, this solution looks nearly like Newtonian gravity with small relativistic corrections. But Schwarzschild noticed something strange in his solution. At a particular radius, now called the Schwarzschild radius, the mathematical description appears to break down. The metric coefficients become infinite or zero, suggesting something dramatic happens there.

For decades, physicists thought this mathematical singularity at the Schwarzschild radius was merely a problem with the chosen coordinate system, an artifact of the mathematics rather than a physical feature. Through the 1950s and 1960s, work by physicists including Martin Kruskal, David Finkelstein, and Roger Penrose clarified the situation. The singularity at the Schwarzschild radius is indeed a coordinate singularity, removable by changing coordinates, but it represents something profoundly physical: an event horizon, a one way membrane in spacetime.

The Schwarzschild radius depends on the mass of the object. For the Sun, it is about three kilometers. For Earth, it is about nine millimeters. These radii are far smaller than the actual sizes of these objects, so the Schwarzschild radius lies deep inside where the spherical symmetry assumption breaks down anyway. But if you could somehow compress the Sun to a sphere smaller than three kilometers radius, or Earth to smaller than nine millimeters radius, something remarkable would happen. The object would form an event horizon at the Schwarzschild radius, creating what we now call a black hole.

An event horizon is a boundary in spacetime separating regions that can communicate with the outside universe from regions that cannot. Outside the event horizon, you can still escape if you accelerate hard enough. At the event horizon itself, you would need to travel at exactly the speed of light to escape. Inside the event horizon, no amount of acceleration in any direction allows escape. All future directed paths through spacetime, even for light, lead inward to the center.

This is not because of a force pulling you inward. It is because spacetime is so severely curved inside the event horizon that the direction we normally think of as "outward in space" has become "forward in time." Just as you cannot travel backward in time under normal circumstances, you cannot travel outward in space inside an event horizon. Moving inward toward the singularity is as inevitable as moving forward in time.

What happens at the center of a black hole? According to general relativity, spacetime curvature becomes infinite at a central singularity. All the mass that formed the black hole, plus anything that has fallen in since, is compressed to infinite density at a single point. Time and space lose their usual meanings. The equations of general relativity break down completely.

Most physicists believe this singularity indicates the limits of general relativity, not an accurate description of reality. A quantum theory of gravity should describe what actually happens at the center of black holes. Perhaps quantum effects prevent infinite densities, creating instead a region of extremely high but finite curvature. Perhaps singularities do not form at all in a quantum gravity theory. We do not know because we do not yet have a complete, experimentally verified theory of quantum gravity.

Black holes can form in several ways. The most common mechanism is stellar collapse. When a massive star exhausts its nuclear fuel, it can no longer support itself against gravitational collapse. If the star's core is more than about three times the Sun's mass, not even neutron degeneracy pressure can halt the collapse. The core implodes, and if enough mass remains, it collapses through its Schwarzschild radius, forming a black hole event horizon.

The formation process is dramatic. The collapsing star's outer layers rebound off the ultra dense core and explode outward in a supernova, one of the most energetic events in the universe. The core continues collapsing inward. From the perspective of a distant observer, the collapse appears to slow as the surface approaches the event horizon due to gravitational time dilation. Light from the surface becomes increasingly redshifted. The star appears to freeze just outside its Schwarzschild radius, growing dimmer and redder but never quite vanishing.

From the perspective of someone falling with the collapsing star, nothing special happens at the event horizon. There is no barrier, no wall, nothing marking the point of no return. You cross the event horizon in finite time on your own clock and continue falling inward toward the singularity. Only later, when you try to send signals outward or turn around, do you discover that escape has become impossible.

This difference between perspectives, one of general relativity's most counterintuitive features, arises from the extreme curvature of spacetime near the event horizon. Time flows at vastly different rates for observers at different positions. What takes a finite time for an infalling observer appears to take infinite time to a distant observer. Both perspectives are correct in their own reference frames.

Rotating black holes are described by the Kerr metric, discovered by Roy Kerr in 1963. Rotation adds complexity to black hole structure. A rotating black hole has two event horizons, an outer and an inner, and a region called the ergosphere outside the outer horizon where spacetime is dragged around by the black hole's rotation. Objects in the ergosphere must rotate in the same direction as the black hole, unable to remain stationary relative to distant stars. The ergosphere allows a process called the Penrose mechanism where energy can be extracted from a rotating black hole, slowing its rotation.

Inside a rotating black hole, the singularity is not a point but a ring, a loop of infinite curvature. The mathematical structure suggests exotic possibilities including paths that avoid the singularity entirely, perhaps exiting through another region of spacetime. These paths would represent shortcuts through spacetime, wormholes connecting distant regions. However, these wormhole solutions typically require traveling through regions of exotic matter or involve naked singularities, suggesting they may not be physically realizable.

Astronomers have identified numerous black hole candidates through their gravitational effects on nearby matter. When a black hole is part of a binary system with a normal star, it can pull matter from the companion star. This matter spirals inward, forming an accretion disk that heats to millions of degrees and emits intense X-rays before crossing the event horizon. The X-ray binary system Cygnus X-1, discovered in 1964, was the first strong black hole candidate and is now confirmed to contain a black hole of about 15 solar masses.

Supermassive black holes, millions to billions of times the Sun's mass, lurk at the centers of most galaxies. The Milky Way's center hosts Sagittarius A star, a black hole of about 4 million solar masses. Observations of stars orbiting this black hole, following Keplerian orbits around an invisible object, provided compelling evidence for its existence. In 2019, the Event Horizon Telescope collaboration released the first direct image of a black hole's shadow, showing the supermassive black hole at the center of the galaxy M87, about 55 million light years away.

How supermassive black holes formed remains uncertain. They were already present in galaxies less than a billion years after the Big Bang, perhaps growing from smaller black holes that merged and accreted matter, or perhaps forming through direct collapse of massive gas clouds in the early universe, or through other exotic formation channels. Their role in galaxy formation and evolution is profound, with black hole feedback from their accretion processes regulating star formation in surrounding regions.

Hawking radiation, predicted by Stephen Hawking in 1974, shows that black holes are not completely black. Quantum effects near the event horizon allow particle antiparticle pairs to separate, with one particle falling into the black hole and the other escaping to infinity. To a distant observer, the black hole appears to emit thermal radiation with a temperature inversely proportional to its mass. For stellar mass black holes, this temperature is tiny, only billionths of a degree above absolute zero. But for microscopic black holes, if they exist, the temperature could be high enough to make them glow and eventually evaporate completely.

This radiation creates a profound puzzle called the black hole information paradox. Quantum mechanics requires that information is never destroyed. But if a black hole evaporates completely through Hawking radiation, which is thermal and random, what happens to the information about everything that fell into the black hole? The radiation carries almost no information about the black hole's formation or contents beyond its mass, charge, and angular momentum. Where does the information go?

Proposed resolutions include the idea that information is encoded in subtle correlations in the Hawking radiation, or that information is preserved at the event horizon in a two dimensional holographic encoding, or that black hole remnants remain after evaporation carrying the information. The paradox touches deep issues about the nature of quantum mechanics, gravity, and spacetime. Most physicists believe information is preserved somehow, but the mechanism remains debated.

Wormholes, hypothetical tunnels through spacetime connecting distant regions, appear in some solutions to Einstein's equations. The simplest wormhole solution, the Einstein Rosen bridge, connects two regions of spacetime through a throat that pinches off so quickly that nothing can traverse it before it closes. To hold a wormhole open long enough for objects to pass through requires exotic matter with negative energy density, violating the energy conditions that normal matter satisfies.

Whether exotic matter exists is unknown. Quantum fields can violate energy conditions in limited regions, allowing temporary negative energy densities. Perhaps a sufficiently advanced civilization could manipulate quantum fields to stabilize a traversable wormhole. Alternatively, wormholes might be fundamentally unstable, collapsing immediately upon formation, making them useless for travel or communication.

If stable wormholes exist, they raise troubling possibilities for causality. A wormhole connecting two points in space could in principle be manipulated to create a time machine, allowing travel to the past. One mouth of the wormhole could be accelerated to high speed and returned, creating a time difference between the mouths due to time dilation. Entering one mouth could then lead to emerging from the other mouth at an earlier time. Such closed timelike curves permit all the paradoxes of time travel, like killing your grandfather before your parent is conceived.

Stephen Hawking proposed the chronology protection conjecture, suggesting that the laws of physics prevent the formation of closed timelike curves. Perhaps quantum effects destroy any configuration that would permit time travel to the past, or perhaps the accumulated vacuum fluctuations create infinite energy density that destroys the time machine at the moment it forms. These ideas remain speculative, but most physicists doubt that nature permits macroscopic time travel to the past.

Black holes and wormholes represent the most extreme predictions of general relativity, regions where spacetime curvature becomes so severe that our intuitions completely fail. Event horizons are one way boundaries in spacetime. Singularities are places where spacetime itself breaks down. Closed timelike curves threaten causality. These exotic structures show that reality can be far stranger than anything in science fiction, with possibilities limited only by fundamental physical principles.

The study of black holes has revealed deep connections between gravity, quantum mechanics, thermodynamics, and information theory. Black holes have entropy proportional to their event horizon surface area, suggesting that all the information about a black hole is encoded holographically on its two dimensional surface. This holographic principle has profound implications for quantum gravity, suggesting that the three dimensional physics we observe might emerge from more fundamental two dimensional physics at the boundary of spacetime.

Black holes are not just theoretical curiosities. They are real objects in our universe, detected through their gravitational effects, their accretion disk emissions, their distortion of spacetime measured through gravitational lensing, and now through direct imaging of their event horizon shadows and gravitational wave signals from merging black holes. Every observation confirms general relativity's predictions about these extreme objects where spacetime curvature dominates.

In black holes, physics meets its limits. General relativity predicts singularities but cannot describe them. The event horizon shields these singularities from observation, but quantum effects suggest information is preserved somehow. The resolution of these puzzles awaits a quantum theory of gravity that can describe spacetime at the Planck scale where quantum effects become dominant. Until then, black holes remain the most mysterious objects in the universe, where our current theories reach their boundaries and the next theory of physics must take over.

III

III. The Universe at Large

Having explored the quantum strange and the relativistic bizarre, we now expand our view to the largest scales imaginable: the universe itself. This part examines the origin, evolution, structure, and composition of the cosmos. We will discover that the universe had a definite beginning, that it continues to expand and accelerate, and that most of its contents remain mysterious, invisible to our telescopes but detectable through gravitational effects.

9. The Big Bang and Cosmic Expansion - How We Know the Universe Had a Beginning

For most of human history, people believed the universe was eternal and unchanging on cosmic scales. Stars might be born and die, planets might orbit, but the overall structure of the cosmos was thought to persist forever, having no beginning and no end. This eternal universe seemed philosophically appealing, avoiding questions about first causes and creation.

In the early twentieth century, this eternal picture collapsed under the weight of astronomical observations and theoretical developments. The universe, it turned out, is not eternal and static but dynamic and evolving, expanding outward from a hot dense beginning about 13.8 billion years ago. This discovery ranks among the most profound in human intellectual history, revealing that time itself had a beginning and that the universe has a definite age and history.

The first hints came from observations of galaxy spectra. In the 1910s and 1920s, astronomers including Vesto Slipher measured the light from distant galaxies using spectroscopy, splitting the light into its component wavelengths to identify characteristic spectral lines from hydrogen and other elements. Nearly all galaxies showed spectral lines shifted toward longer wavelengths compared to laboratory measurements. This redshift indicated that galaxies were moving away from Earth.

In 1929, Edwin Hubble combined these velocity measurements with estimates of galaxy distances based on observations of variable stars and other distance indicators. He discovered a linear relationship: more distant galaxies are receding faster. Specifically, recession velocity is proportional to distance, with the proportionality constant now called the Hubble constant. This relationship, Hubble's law, holds for all sufficiently distant galaxies in all directions.

This universal recession does not mean Earth occupies a special location. Rather, space itself is expanding, carrying galaxies apart. Imagine dots painted on a balloon's surface. As you inflate the balloon, every dot sees every other dot moving away, with more distant dots receding faster. No dot occupies a center; the expansion is the same from every dot's perspective. Similarly, every galaxy sees all other galaxies receding, with recession velocity proportional to distance.

If space is expanding, it must have been denser in the past. Running the expansion backward in time, galaxies were closer together, the universe was hotter and denser, and eventually, at a finite time in the past, everything was compressed to infinite density. This moment of infinite density is the Big Bang singularity, the apparent beginning of space, time, matter, and energy.

The term "Big Bang" was actually coined as a dismissive label by astronomer Fred Hoyle, a proponent of the competing steady state theory who did not believe the universe had a beginning. But observations increasingly favored the Big Bang model, and the term stuck, though it is somewhat misleading. The Big Bang was not an explosion of matter outward into pre existing empty space. It was the beginning of space itself, happening everywhere simultaneously, with no center and no edge.

One might ask what happened before the Big Bang or what caused it. These questions may be meaningless. If time itself began at the Big Bang, there was no "before." Asking what happened before the Big Bang is like asking what is north of the North Pole. The question presupposes a framework, time extending to the past infinitely or space extending beyond Earth's poles, that does not exist. Time itself is part of the universe and came into existence with it.

General relativity, applied to a homogeneous expanding universe, yields the Friedmann equations describing cosmic evolution. These equations relate the expansion rate to the density and pressure of the universe's contents. For a universe containing ordinary matter and radiation, the equations predict that expansion should decelerate over time due to gravitational attraction pulling everything together. The universe might expand forever, gradually slowing but never stopping, or it might eventually recollapse in a Big Crunch, depending on the total density.

Evidence for the Big Bang comes from multiple independent sources beyond Hubble's law. The most compelling is the cosmic microwave background, or CMB, discovered accidentally in 1964 by Arno Penzias and Robert Wilson. They were testing a sensitive radio antenna for Bell Labs when they detected a persistent background noise at microwave frequencies coming uniformly from all directions. After ruling out terrestrial sources, including famously evicting pigeons nesting in the antenna, they realized the signal was cosmic in origin.

The CMB is radiation from the early universe when it was hot enough, about 3000 Kelvin, that matter existed as ionized plasma. Photons scattered constantly off free electrons, keeping matter and radiation in thermal equilibrium. As the universe expanded and cooled, electrons combined with nuclei to form neutral atoms in an event called recombination, occurring about 380,000 years after the Big Bang. Once atoms formed, photons could travel freely without scattering. We observe these photons today, redshifted by the expansion of space from visible light to microwave radiation.

The CMB has almost perfect blackbody spectrum, exactly as predicted for radiation from a hot dense early universe. Its temperature is extraordinarily uniform, about 2.725 Kelvin in all directions, varying by only about one part in 100,000. This uniformity shows the early universe was remarkably smooth, with nearly identical density everywhere. The tiny temperature variations, first detected by the COBE satellite in 1992 and measured precisely by later missions including WMAP and Planck, represent primordial density fluctuations that would later grow through gravitational amplification into galaxies and galaxy clusters.

The pattern of these temperature fluctuations provides a wealth of information about cosmological parameters. The angular size of the largest fluctuations depends on the universe's geometry. The relative heights of fluctuations at different scales depend on the densities of ordinary matter, dark matter, and dark energy. From CMB data alone, cosmologists can determine the universe's age, composition, and geometry with remarkable precision.

Big Bang nucleosynthesis provides another independent confirmation. In the first few minutes after the Big Bang, when the universe was hot enough for nuclear reactions but cool enough that deuterium was not immediately photodisintegrated, fusion reactions converted about 25 percent of the hydrogen into helium, with trace amounts of deuterium, tritium, lithium, and beryllium. The predicted abundances depend sensitively on the density of ordinary baryonic matter. Observations of primordial helium and deuterium abundances in very old stars and intergalactic gas match Big Bang predictions precisely, confirming both the Big Bang framework and the density of ordinary matter.

The abundances also show that ordinary atomic matter can account for only about 5 percent of the universe's total energy density. Additional evidence from galaxy rotation curves, gravitational lensing, and large scale structure formation indicates that about 27 percent of the universe consists of dark matter, some form of matter that does not emit or absorb light but does exert gravitational influence. The remaining 68 percent consists of dark energy, causing the expansion to accelerate rather than decelerate.

The timeline of the Big Bang begins at the Planck time, about 10 to the negative 43 seconds after the singularity, when quantum gravitational effects were dominant and our current theories break down. We cannot describe physics before this time without a quantum theory of gravity. After the Planck time, the universe underwent a period of exponential expansion called inflation, during which the universe doubled in size at least 60 times in a tiny fraction of a second. Inflation explains why the universe is so uniform, why it appears geometrically flat, and why we do not observe magnetic monopoles or other exotic relics predicted by particle physics.

After inflation ended, the universe was filled with a hot soup of fundamental particles and antiparticles constantly being created and annihilated. As the universe expanded and cooled, it underwent a series of phase transitions. At about one microsecond after the Big Bang, quarks combined to form protons and neutrons. At about one second, the temperature dropped below the threshold for creating electron positron pairs, and most antimatter annihilated with matter, leaving a small excess of matter that comprises everything we observe today.

From one second to three minutes after the Big Bang, Big Bang nucleosynthesis occurred, fusing protons and neutrons into light element nuclei. At about 380,000 years, recombination occurred and the universe became transparent to light, releasing the cosmic microwave background radiation we observe today. For the next few hundred million years, the universe was dark, with no stars yet formed, a period called the cosmic dark ages.

The first stars formed from primordial hydrogen and helium compressed by gravity into dense clouds that ignited nuclear fusion. These first stars were likely massive, short lived, and contained no elements heavier than lithium. When they exploded as supernovae, they seeded the universe with heavier elements created in their cores and during their explosions. Subsequent generations of stars formed from gas enriched with these heavy elements, allowing the formation of planets and eventually life.

Galaxies assembled through gravitational amplification of primordial density fluctuations. Small density enhancements attracted more matter, growing into larger structures. Dark matter, which interacts only gravitationally, formed the scaffolding on which galaxies assembled, with ordinary matter falling into dark matter halos and forming visible galaxies. Over billions of years, galaxies merged and evolved, forming the large scale structure we observe today, a cosmic web of galaxy clusters, filaments, and voids.

The universe's expansion rate has changed over time. In the early universe dominated by radiation, expansion decelerated rapidly. As matter came to dominate, deceleration continued but more gradually. Then, about 5 billion years ago, dark energy became dominant and expansion began accelerating. The universe will continue expanding and accelerating forever, growing ever larger and colder, with galaxies beyond our local group eventually receding faster than light, disappearing beyond the cosmic horizon.

This timeline, from the Big Bang to the present and into the future, is supported by converging lines of evidence from different areas of physics and astronomy. The expansion of space, the cosmic microwave background, Big Bang nucleosynthesis, the formation and evolution of galaxies, the large scale structure of the universe, and the accelerating expansion all fit together into a coherent picture of cosmic history.

Some questions remain open. What caused the Big Bang? Why did inflation occur? What is dark matter made of? What is dark energy? Is our universe the only one, or part of a vast multiverse? These unanswered questions drive ongoing research, but the basic framework of Big Bang cosmology is as well established as any major theory in physics.

The realization that the universe had a beginning represents a profound shift in human understanding. Time is not eternal. The cosmos has a history. Everything we observe, from atoms to galaxies, emerged from an inconceivably hot dense state 13.8 billion years ago and evolved through physical processes we can study and understand. The universe's story is written in the light from distant galaxies, the cosmic microwave background, the abundances of elements, and the distribution of matter across cosmic scales.

We live in a universe that began in a Big Bang, expanded and cooled, formed atoms and stars and galaxies, and continues expanding and accelerating into an uncertain future. This is not speculation or philosophy. It is the conclusion drawn from decades of careful observations and theoretical work, confirmed by multiple independent lines of evidence. The Big Bang is as well established as any scientific theory describing events beyond direct human observation.

10. Dark Matter - The Invisible Scaffolding of Galaxies

In the 1930s, astronomer Fritz Zwicky studied the Coma galaxy cluster, a collection of over a thousand galaxies bound together by gravity. He measured the velocities of individual galaxies within the cluster using the Doppler shift of their spectral lines. Then he estimated the cluster's total mass based on how much visible matter the galaxies contained. He found a disturbing discrepancy.

The galaxies were moving far too fast. According to his mass estimate from visible matter, the cluster's gravity was far too weak to hold such fast moving galaxies. They should have escaped the cluster long ago, scattering into intergalactic space. Yet the cluster clearly existed and appeared stable. Zwicky concluded that the cluster must contain far more mass than he could see, perhaps 400 times more dark matter than luminous matter, providing the extra gravity needed to bind the cluster.

For decades, most astronomers dismissed Zwicky's idea. Perhaps his mass estimates were wrong. Perhaps his velocity measurements were flawed. Perhaps the cluster was not in equilibrium but recently formed or in the process of dispersing. Dark matter seemed too radical, too speculative, too contrary to the straightforward assumption that what you see is what you get.

Then in the 1970s, astronomer Vera Rubin performed detailed studies of galaxy rotation curves. She measured how fast stars and gas orbit around galactic centers at various distances from the center. According to Newton's laws, orbital velocity should decrease with distance from the center, just as outer planets orbit the Sun more slowly than inner planets. The visible matter in galaxies is concentrated toward the center, so stars far from the center should orbit slowly.

Rubin found the opposite. Stars far from galactic centers orbit just as fast as stars closer in, with rotation curves remaining flat rather than declining with distance. This flat rotation curve indicated that galaxies contain large amounts of invisible mass extending far beyond the visible disk, with the total mass increasing with radius even where no stars or gas are visible. The distribution of this dark matter forms an extended halo surrounding each galaxy.

Additional evidence accumulated from other observations. Gravitational lensing, the bending of light by massive objects predicted by general relativity, showed that galaxy clusters contain far more mass than visible matter can account for. The deflection of light from background galaxies passing near clusters indicated total masses five to ten times larger than the mass in stars and gas. The cosmic microwave background temperature fluctuations showed patterns consistent with a universe containing about five times more dark matter than ordinary matter. Large scale structure simulations demonstrated that galaxies and galaxy clusters could not have formed in the time available without dark matter providing extra gravitational attraction to amplify density fluctuations.

Today, the evidence for dark matter is overwhelming and comes from multiple independent sources. Galaxy rotation curves, galaxy cluster dynamics, gravitational lensing, cosmic microwave background anisotropies, Big Bang nucleosynthesis, and large scale structure formation all point to the same conclusion: about 85 percent of the matter in the universe is dark matter that does not emit, absorb, or reflect electromagnetic radiation. We detect it only through its gravitational effects.

What is dark matter made of? The answer remains one of the biggest mysteries in physics. We know what it is not. It is not ordinary matter that happens to be dark, like black holes, brown dwarfs, or cold gas clouds. Big Bang nucleosynthesis constrains the density of ordinary atomic matter to about 5 percent of the universe's total energy density, far less than the 27 percent attributed to dark matter. Dark matter must be something exotic, some form of matter not composed of protons, neutrons, and electrons.

The leading candidate is weakly interacting massive particles, or WIMPs. These hypothetical particles would interact only through gravity and possibly the weak nuclear force, not through electromagnetism or the strong force. They would have been produced in the early universe in just the right abundance to account for observed dark matter density. Many theories of particle physics beyond the Standard Model predict particles with these properties, including supersymmetry which posits that every Standard Model particle has a heavier supersymmetric partner.

Despite decades of searching, no one has detected WIMPs directly. Experiments deep underground use ultra sensitive detectors shielded from cosmic rays, looking for rare interactions between dark matter particles and atomic nuclei. The idea is that as Earth moves through the dark matter halo surrounding our galaxy, dark matter particles should occasionally collide with detector nuclei, producing detectable recoils. So far, these experiments have found nothing conclusive, placing increasingly stringent limits on WIMP properties.

Another possibility is axions, extremely light particles predicted by some theories addressing problems in quantum chromodynamics. Axions would be produced abundantly in the early universe and could comprise dark matter if their mass is in the right range. Axion detection experiments search for the conversion of axions to photons in strong magnetic fields. Like WIMP searches, axion experiments have not yet found definitive signals but continue to improve in sensitivity.

More exotic candidates include primordial black holes formed in the early universe, sterile neutrinos that interact even more weakly than ordinary neutrinos, or particles from completely unexpected new physics. The diversity of candidates reflects our ignorance. We know dark matter exists, we know roughly how much there is, we know it is cold, meaning non relativistic during structure formation, but we do not know what it is.

Dark matter plays a crucial role in cosmic structure formation. In the early universe, ordinary matter and dark matter were both distributed nearly uniformly with tiny density fluctuations. As the universe expanded, these fluctuations grew through gravitational instability. Regions slightly denser than average attracted more matter, becoming denser still, eventually collapsing into gravitationally bound structures.

Ordinary matter would have taken too long to form galaxies if it were the only matter present. Radiation pressure from the cosmic microwave background and interactions between matter and radiation would have smoothed out density fluctuations in ordinary matter until recombination. Without dark matter, galaxies would only now be beginning to form, and the universe would look very different.

Dark matter, unaffected by radiation pressure because it does not interact electromagnetically, began collapsing into structures much earlier. It formed the gravitational potential wells into which ordinary matter fell after recombination. Galaxies formed within dark matter halos, with stars and gas condensing at the centers of these invisible scaffolds. The dark matter halo provides most of the gravitational mass holding galaxies together and determining their dynamics.

Computer simulations of structure formation starting from primordial density fluctuations and evolving forward billions of years reproduce the observed large scale structure of the universe only when dark matter is included. Without dark matter, simulations produce far too little structure. With dark matter in the amounts indicated by observations, simulations reproduce the cosmic web of filaments, clusters, and voids we observe.

These simulations also predict the internal structure of dark matter halos. They should have dense cores and extended outer regions, with density decreasing with distance from the center following specific profiles. Observations of galaxy rotation curves and gravitational lensing generally confirm these predictions, though some discrepancies remain on small scales where baryonic physics effects from star formation and feedback may complicate the picture.

One ongoing puzzle is the apparent absence of dark matter in some galaxies. Certain ultra diffuse galaxies appear to contain little or no dark matter, with their stars' motions explained entirely by the mass of visible stars and gas. If dark matter exists, why do these galaxies lack it? Perhaps dark matter was stripped away by interactions with other galaxies, or perhaps these galaxies formed through unusual processes that expelled dark matter. Alternatively, they might challenge our understanding of dark matter itself, suggesting more complex behavior than simple collision less particles.

Modified Newtonian Dynamics, or MOND, proposes an alternative explanation for galactic rotation curves and other dark matter evidence. MOND suggests that Newton's laws break down at very low accelerations, with gravity behaving differently in weak gravitational fields like galaxy outskirts. Modified gravity could potentially explain rotation curves without dark matter. However, MOND struggles to explain other dark matter evidence like the Bullet Cluster, where two galaxy clusters collided and the gravitational lensing signal is displaced from the visible matter, showing that most of the mass is concentrated where dark matter would be if it passed through unimpeded, while ordinary matter interacted and was slowed. MOND also cannot easily explain the cosmic microwave background or structure formation.

Most physicists favor dark matter over modified gravity, but the lack of direct detection keeps the question open. Perhaps dark matter is more complicated than simple WIMPs or axions. Perhaps it interacts more richly than through gravity alone, forming dark atoms or dark molecules. Perhaps it decays slowly, producing faint signals we have not yet detected. Perhaps multiple types of dark matter exist, each contributing to the total dark matter density.

The search for dark matter continues on multiple fronts. Direct detection experiments grow ever more sensitive, capable of detecting increasingly rare and weak interactions. Indirect detection experiments search for signals from dark matter annihilation or decay, looking for excess gamma rays, cosmic rays, or neutrinos from regions where dark matter should be concentrated. Collider experiments at the Large Hadron Collider search for dark matter particles produced in high energy collisions, hoping to create them in the laboratory and study their properties directly.

Identifying dark matter would rank among the most important discoveries in physics. It would reveal new fundamental particles beyond the Standard Model, potentially opening windows into deeper theories of physics. It would confirm our understanding of cosmic structure formation and galaxy dynamics. It would answer one of the most basic questions about the universe: what is most of the matter made of?

For now, dark matter remains mysterious, detected through its gravitational effects but invisible to our telescopes, dominating the matter content of the universe yet unidentified in our laboratories. It forms the scaffolding on which galaxies assembled and holds them together today. It shapes the large scale structure of the cosmos and played a crucial role in creating the universe we observe. Yet we do not know what it is, a humbling reminder that despite our progress in understanding reality, the universe still holds profound secrets.

11. Dark Energy - Why the Universe Is Accelerating Apart

In 1998, two independent teams of astronomers studying distant supernovae made a discovery that shocked the physics community. They were measuring the distances and recession velocities of Type Ia supernovae, exploding white dwarf stars that serve as standard candles because of their consistent peak brightness. By comparing the apparent brightness of these supernovae to their known intrinsic brightness, the teams could determine their distances. By measuring the Redshift of their host galaxies, they could determine their recession velocities.

The expectation was that distant supernovae would be slightly brighter than predicted by a steadily expanding universe because gravity should slow the expansion over time. The universe's expansion rate should have been faster in the past, meaning distant supernovae should be closer than they would be in a universe expanding at a constant rate. Instead, the teams found the opposite. Distant supernovae were fainter than expected, indicating they were farther away than predicted. The universe's expansion was not slowing down but speeding up.

This acceleration defied intuition. Gravity attracts. Every piece of matter in the universe pulls on every other piece. This mutual gravitational attraction should slow the expansion, perhaps eventually halting it and causing recollapse, or at least causing the expansion rate to decrease over time. Yet observations showed the expansion accelerating, as if some repulsive force were pushing space apart.

To explain this acceleration, physicists invoked dark energy, a mysterious component filling all of space and exerting negative pressure that drives accelerated expansion. Unlike dark matter, which clumps gravitationally like ordinary matter but happens to be invisible, dark energy appears uniformly distributed throughout space with constant density even as the universe expands. About 68 percent of the universe's total energy density consists of dark energy.

The simplest candidate for dark energy is the cosmological constant, a term Einstein added to his field equations in 1917 to allow static universe solutions. When he learned the universe was expanding, Einstein abandoned the cosmological constant, calling it his greatest blunder. But the cosmological constant has made a comeback as the leading explanation for dark energy.

A cosmological constant represents the energy of empty space itself, vacuum energy that exists even when no matter or radiation is present. Quantum field theory predicts that empty space is not truly empty but filled with quantum fluctuations, virtual particles constantly appearing and disappearing. These fluctuations contribute energy density to the vacuum, and this energy density acts like a cosmological constant.

The problem is that naive calculations of vacuum energy density using quantum field theory give answers about 120 orders of magnitude larger than the observed dark energy density. This cosmological constant problem is one of the worst fine tuning problems in physics. Why is the vacuum energy so small compared to quantum field theory predictions? No one knows. Somehow, contributions from different fields must cancel to extraordinary precision, leaving only a tiny residual energy density that manifests as dark energy.

Despite this theoretical embarrassment, the cosmological constant fits observational data well. If dark energy density is constant, neither increasing nor decreasing as the universe expands, and if its equation of state parameter w equals negative one, meaning its pressure equals negative one times its energy density, then observations are well explained. More complex models where dark energy density changes over time or w differs from negative one are allowed by current data but not required.

Alternative explanations for cosmic acceleration have been proposed. Quintessence models invoke a dynamical scalar field rolling down a potential, similar to the inflaton field thought to drive inflation in the early universe. Unlike a cosmological constant, quintessence energy density can change over time and w can differ from negative one. Quintessence requires introducing new fields and fine tuning their properties, but it avoids the extreme fine tuning of the cosmological constant itself.

Modified gravity theories suggest that general relativity breaks down on cosmological scales and requires modification. Perhaps gravity becomes repulsive at very large distances, or perhaps additional dimensions of space affect gravitational dynamics on cosmic scales. These models face challenges explaining the full range of cosmological observations and often require their own fine tuning to match data.

The physics community remains divided about which explanation is correct. Some favor the cosmological constant as the simplest option consistent with data. Others pursue quintessence or modified gravity hoping to find deeper explanations. Still others suspect the answer involves new physics we have not yet imagined, perhaps connected to quantum gravity or string theory.

Dark energy has profound implications for the universe's future. With dark energy dominating and driving accelerated expansion, the universe will expand forever, growing larger and colder with time. Galaxies beyond our local group, already receding from us, will accelerate away ever faster. Eventually, they will recede faster than light, not because they move through space faster than light but because space itself expands rapidly enough that light from them can never reach us. They will disappear beyond the cosmic horizon.

In the far future, hundreds of billions of years from now, observers in our galaxy will see only the Local Group galaxies gravitationally bound to us. The rest of the universe will have receded beyond the observable horizon. If those future astronomers discovered cosmology then, with no visible evidence of distant galaxies or cosmic expansion, would they ever figure out the universe's history?

Eventually, even stars will exhaust their fuel. The last stars will fade and die, leaving behind white dwarfs, neutron stars, and black holes. These remnants will slowly evaporate through Hawking radiation on timescales of 10 to the 100th power years or more. The universe will approach a state of maximum entropy, a cold dispersed collection of elementary particles and radiation, with no remaining structure or organization.

This ultimate fate depends on the nature of dark energy. If dark energy is a cosmological constant, it remains constant forever and the universe expands eternally. If dark energy increases over time, it could lead to a Big Rip, where accelerating expansion eventually tears apart galaxies, then stars, then atoms. If dark energy decreases and becomes negative, expansion could reverse, leading to a Big Crunch where the universe recollapses. Current observations favor constant dark energy, but we cannot rule out more exotic possibilities.

The accelerating expansion affects cosmic structure formation. Dark matter continues to clump gravitationally, forming larger and larger structures, but dark energy opposes this clumping on the largest scales. Galaxy clusters can still form through gravitational collapse, but superclusters larger than a few hundred million light years cannot form because dark energy's negative pressure prevents such large structures from collapsing. This sets a maximum scale for cosmic structure.

Measuring dark energy's properties is a major goal of modern cosmology. Surveys map the distribution of galaxies across cosmic history, tracing how structure formation evolves over time in the presence of dark energy. Observations of Type Ia supernovae at various distances and redshifts constrain the expansion history. Studies of weak gravitational lensing measure how light from distant galaxies is bent by intervening matter, probing the growth of structure. Measurements of baryon acoustic oscillations, characteristic scales imprinted in galaxy distributions by sound waves in the early universe, provide a standard ruler for cosmology.

All these observations consistently indicate that about 68 percent of the universe consists of dark energy with properties close to a cosmological constant. But the uncertainty is large enough that quintessence models cannot be ruled out. Determining whether dark energy is truly constant or slowly varying requires more precise measurements over larger volumes of space and longer spans of cosmic time.

The combination of dark matter and dark energy means that about 95 percent of the universe's energy density consists of components we do not understand. Ordinary matter, the protons, neutrons, and electrons making up atoms, stars, planets, and people, accounts for only 5 percent. We are made of the minority component, living in a universe dominated by invisible dark matter and mysterious dark energy.

This realization is both humbling and exciting. Humbling because it shows how much we do not know. Exciting because it reveals that profound discoveries await. Understanding dark matter and dark energy will require new physics beyond our current theories. Perhaps supersymmetry or extra dimensions or quantum gravity. Perhaps completely unexpected phenomena that will only become clear with better observations and deeper theoretical work.

For now, we observe the effects of dark energy through cosmic acceleration. We measure its density and equation of state. We constrain how it affects structure formation and cosmic expansion. We do not know what it is, why it has the density it does, or whether it is truly constant. These questions drive some of the most ambitious experiments in physics and astronomy, from ground based telescopes to space satellites to next generation surveys mapping billions of galaxies.

Dark energy stands alongside dark matter as one of the greatest mysteries in modern physics. Together, they dominate the universe's content and shape its past, present, and future. Together, they remind us that despite all we have learned about reality, from quantum mechanics to relativity to the Big Bang, the universe still holds secrets that challenge our understanding and push the boundaries of human knowledge.

IV

IV. What It All Means

Having journeyed through the quantum world, explored relativity and extreme spacetime, and surveyed the cosmos from the Big Bang to dark energy, we arrive at the question of meaning. What does modern physics tell us about reality? How do these strange discoveries connect to everyday life? And what implications do they have for technology, philosophy, and our understanding of existence itself?

12. Quantum Tech in Your Pocket - From Transistors to Quantum Computers

When you use a smartphone, computer, or any electronic device, you are holding applied quantum mechanics in your hand. The transistors forming the billions of logic gates in microprocessors function because of quantum tunneling. The LED screen emits light through quantum electroluminescence. The battery stores energy through quantum chemistry. The GPS satellite signals accounting for relativistic time dilation let you navigate. Modern technology is quantum technology, built on principles that defied common sense when first discovered but now enable the devices we depend on daily.

The transistor, invented in 1947, was the first practical quantum device and remains the most important. Transistors are semiconductor switches that control electrical current. Traditional vacuum tubes performed similar functions but were large, fragile, power hungry, and short lived. Transistors are tiny, robust, efficient, and essentially permanent, allowing them to be miniaturized and integrated by the billions onto single silicon chips.

How transistors work depends entirely on quantum mechanics. Semiconductors like silicon have electrical conductivity between conductors and insulators. In pure silicon, electrons are bound to atoms and cannot move freely. But silicon can be doped with impurity atoms that either donate extra electrons, creating n type semiconductors, or accept electrons, creating holes that act like positive charge carriers, creating p type semiconductors.

A transistor combines n type and p type regions in specific geometries. The most common design, the MOSFET or metal oxide semiconductor field effect transistor, has source and drain regions of one type separated by a channel of the opposite type, with a gate electrode controlling channel conductivity. Applying voltage to the gate changes how easily current flows between source and drain, allowing the transistor to act as an electrically controlled switch or amplifier.

The key quantum effect is band structure. In quantum mechanics, electrons in solids occupy energy bands separated by band gaps where no electron states exist. Semiconductors have small band gaps that can be bridged by thermal energy or doping, allowing controllable conductivity. The detailed behavior of electrons in semiconductors, including how they respond to electric fields and how they tunnel through thin barriers, requires quantum mechanics to describe accurately.

Modern microprocessors contain billions of transistors with features now smaller than 10 nanometers, approaching atomic scales where quantum effects become increasingly pronounced. Engineers must account for quantum tunneling through thin gate oxides, quantum confinement in nanoscale structures, and other quantum phenomena when designing these devices. The entire semiconductor industry rests on quantum mechanical engineering.

Flash memory in smartphones and USB drives stores data by trapping electrons in floating gates through quantum tunneling. When you write data, electrons tunnel through an insulating barrier onto the floating gate. The trapped charge changes the transistor's threshold voltage, encoding a bit of information. When you erase data, electrons tunnel back off the floating gate. Reading does not require tunneling, just sensing whether the threshold voltage is high or low. This tunneling process allows non volatile storage that retains data without power.

LEDs and laser diodes emit light through quantum processes. In an LED, electrons and holes recombine in a semiconductor, releasing energy as photons. The photon energy equals the band gap energy, determining the light's color. Blue LEDs, which enabled white LED lighting and won their inventors the 2014 Nobel Prize, required developing gallium nitride semiconductors with large enough band gaps to emit blue light.

Lasers add stimulated emission. In a laser, photons trigger the emission of additional photons with identical energy, phase, and direction, producing coherent light. This requires a population inversion where more electrons occupy excited states than ground states, achieved by pumping energy into the system. Lasers find applications from barcode scanners to fiber optic communications to laser cutting to medical surgery to scientific research.

Solar cells convert sunlight to electricity through the photovoltaic effect. When photons strike a semiconductor, they can excite electrons across the band gap, creating electron hole pairs. In a solar cell with a p-n junction, the built in electric field separates electrons and holes before they recombine, driving current through an external circuit. The efficiency depends on the band gap matching the solar spectrum, a quantum mechanical optimization problem.

MRI scanners image soft tissues by detecting nuclear magnetic resonance, a purely quantum phenomenon. Atomic nuclei with spin, like hydrogen in water molecules, align with magnetic fields and precess at characteristic frequencies. Radio waves at the resonance frequency flip the spins, and detectors measure the signals emitted as spins relax back to equilibrium. Different tissues have different relaxation times, allowing soft tissue contrast that X-rays cannot provide.

PET scanners detect positron emission from radioactive tracers injected into patients. When a positron encounters an electron, they annihilate, producing two gamma ray photons flying in opposite directions. Detectors surrounding the patient detect these photons and reconstruct where the annihilation occurred, creating three dimensional images of tracer distribution showing metabolic activity, blood flow, or tumor locations.

Atomic clocks, the most precise timekeeping devices ever built, use quantum transitions in atoms as frequency references. Cesium atomic clocks define the second as 9,192,631,770 cycles of radiation corresponding to a specific hyperfine transition in cesium 133. Optical atomic clocks using transitions in ions like strontium or aluminum achieve even higher precision, losing less than one second over the age of the universe. GPS satellites carry atomic clocks, and their signals must account for relativistic time dilation to achieve meter level positioning accuracy.

Quantum cryptography offers theoretically unbreakable secure communication. The most mature implementation, quantum key distribution, uses quantum states of photons to establish shared secret keys between parties. Any eavesdropper attempting to intercept the photons must measure them, disturbing their quantum states due to the uncertainty principle and revealing the presence of eavesdropping. The security comes from quantum mechanics itself, not computational complexity or key secrecy. Quantum cryptography networks are already deployed commercially in some countries.

Quantum computers represent the next frontier of quantum technology. Unlike classical computers that store information in bits that are either zero or one, quantum computers use qubits that can exist in superpositions of zero and one. A system of n qubits can exist in a superposition of all 2 to the n possible combinations simultaneously. Quantum algorithms can exploit this parallelism to solve certain problems exponentially faster than classical algorithms.

The most famous quantum algorithm is Shor's algorithm for factoring large numbers. Factoring is believed to be hard for classical computers, with the best known algorithms requiring exponential time. Shor's algorithm factors in polynomial time, potentially breaking RSA encryption which relies on factoring difficulty. A large scale quantum computer running Shor's algorithm could break much of current internet security, motivating research into post quantum cryptography resistant to quantum attacks.

Grover's algorithm provides quadratic speedup for unstructured search problems. Quantum simulation algorithms can efficiently simulate other quantum systems, potentially revolutionizing drug discovery, materials science, and chemistry by allowing detailed quantum mechanical simulations of molecules and materials impossible to compute classically.

Building quantum computers faces enormous challenges. Qubits must be isolated from their environment to maintain quantum coherence, yet controllable to perform computations. Decoherence from environmental interactions destroys quantum superpositions in microseconds or milliseconds.

Quantum error correction can protect against decoherence by encoding logical qubits in multiple physical qubits, but this requires extremely low error rates and many additional qubits.

Several physical platforms are being developed. Superconducting qubits, used by companies like Google and IBM, employ circuits cooled to millikelvin temperatures where they exhibit quantum behavior. Trapped ion qubits use individual ions held in electromagnetic traps and manipulated with lasers. Topological qubits, still theoretical, would encode quantum information in exotic quasiparticles resistant to local perturbations. Silicon spin qubits might eventually be integrated with conventional semiconductor manufacturing.

Current quantum computers contain dozens to hundreds of qubits, enough for proof of concept demonstrations and small scale quantum simulations but far from the millions of qubits needed for practical factoring or quantum simulation of complex systems. Achieving fault tolerant quantum computation at scale remains a grand challenge, requiring advances in qubit quality, error correction, control systems, and algorithms.

Beyond these practical applications, quantum phenomena pervade everyday life in less obvious ways. The stability of atoms and molecules depends on quantum mechanics. Chemical bonds are quantum mechanical. The properties of materials from conductivity to magnetism to superconductivity arise from quantum effects in solids. Biology relies on quantum chemistry for enzyme catalysis, photosynthesis, and molecular recognition. Vision begins with quantum photon absorption in retinal molecules. Smell may involve quantum tunneling in olfactory receptors.

Even phenomena we consider purely classical often have quantum origins. The rigidity of matter, the fact that your hand does not pass through tables, comes from the Pauli exclusion principle preventing electron wavefunctions from overlapping. The stability of stars and planets depends on quantum degeneracy pressure. The fusion reactions powering the Sun require quantum tunneling for nuclei to overcome electromagnetic repulsion.

Modern physics is not separate from everyday reality but constitutes the deep explanation of everyday reality. The reason things behave as they do, from the smallest scales to the largest, is quantum mechanics and relativity.

Technology makes this abstract physics concrete and useful. Every electronic device, every medical scanner, every atomic clock is a testament to how well we understand quantum mechanics and how effectively we can engineer quantum systems.

The progression from discovering quantum effects in the early twentieth century to building quantum computers in the early twenty first century demonstrates the practical power of fundamental physics. Understanding nature at its deepest levels enables technologies that transform society. The next generation of quantum technologies, including quantum computers, quantum sensors, and quantum communication networks, will likely revolutionize computing, security, medicine, and scientific research as profoundly as transistors revolutionized the twentieth century.

Quantum mechanics began as a bizarre, counterintuitive theory forced on reluctant physicists by stubborn experimental facts. It became the most precisely tested and accurately confirmed theory in history. And it evolved into the foundation of modern technology. This is the ultimate validation of physics as a way of understanding reality: theories developed to explain abstract experiments eventually enable concrete technologies that change the world.

13. The Quest for the Theory of Everything - Unifying Quantum and Gravity

Physics has achieved extraordinary success in describing the natural world. Quantum mechanics explains atoms, molecules, solids, and fundamental particles with extraordinary precision. General relativity describes gravity, spacetime, black holes, and cosmic expansion beautifully. Together with the Standard Model of particle physics, these theories account for almost all observed phenomena from the smallest subatomic scales to the largest cosmic structures.

Yet these two great theories are fundamentally incompatible. Quantum mechanics treats space and time as a fixed background on which quantum fields evolve. General relativity treats spacetime as a dynamical entity that curves and flexes in response to matter and energy. Quantum mechanics is built on uncertainty and probabilistic outcomes. General relativity is deterministic, with spacetime geometry fully determined by mass energy distribution. Quantum mechanics works with discrete quanta and finite numbers of degrees of freedom. General relativity describes continuous fields with infinite degrees of freedom.

This incompatibility becomes critical in extreme conditions where both theories should apply. At the center of black holes, spacetime curvature becomes infinite according to general relativity, but quantum effects should become important at such extreme densities. In the first instants after the Big Bang, the entire universe was compressed to subatomic scales where quantum effects dominated, yet spacetime was also extremely curved. Understanding these situations requires a quantum theory of gravity, a unified framework combining quantum mechanics and general relativity.

The search for this unified theory has consumed theoretical physics for decades. It represents the holy grail of fundamental physics, the theory that would explain all forces and particles within a single mathematical framework. Several approaches have been pursued, each with successes and challenges, none yet experimentally confirmed.

String theory proposes that fundamental particles are not point like objects but tiny one dimensional strings vibrating in ten dimensional spacetime. Different vibration modes of the string correspond to different particles. A string vibrating one way appears as an electron, vibrating another way as a photon, and so forth. Crucially, one vibration mode corresponds to the graviton, the hypothetical quantum particle of gravity, naturally incorporating gravity into the quantum framework.

String theory requires six extra dimensions beyond the familiar three spatial dimensions and one time dimension. These extra dimensions must be compactified, curled up so small that we do not observe them directly. The geometry of these compactified dimensions determines the properties of particles and forces in the familiar four dimensions, potentially explaining features of the Standard Model like particle masses and force strengths.

String theory has several remarkable mathematical properties. It is anomaly free, avoiding mathematical inconsistencies that plague other attempts to quantize gravity. It naturally includes gauge symmetries underlying the Standard Model forces. It predicts supersymmetry, a symmetry relating bosons and fermions that could explain dark matter and solve other problems in particle physics. It suggests a holographic principle where physics in some region of spacetime can be described by a theory on the boundary of that region.

However, string theory faces serious challenges. It predicts far more possible universes than our single observed universe, perhaps 10 to the 500 power different vacuum states corresponding to different compactifications of the extra dimensions. This vast landscape of possibilities makes it difficult to extract definite predictions. The theory requires energies far beyond current experiments to test directly. Critics argue that string theory is more a mathematical framework than a falsifiable physical theory.

Loop quantum gravity takes a different approach. Rather than replacing particles with strings, it quantizes spacetime itself. In loop quantum gravity, space is not continuous but composed of discrete chunks, quantum geometries that come in fundamental units. Spacetime is woven from loops and nodes forming a spin network, with geometry emerging from the relationships between these fundamental structures.

Loop quantum gravity naturally leads to a discrete quantum structure for space. Area and volume become quantized, coming in discrete units rather than continuous values. The Planck length, about 10 to the negative 35 meters, sets the fundamental scale below which spacetime loses its continuous character. At this scale, the smooth geometry of general relativity breaks down and quantum geometry takes over.

One notable success of loop quantum gravity is resolving the black hole information paradox, at least in principle. By quantizing geometry at the event horizon, loop quantum gravity calculates black hole entropy from counting quantum states of the horizon geometry, reproducing the Bekenstein Hawking formula. It also suggests that the singularity at a black hole's center might be replaced by a region of extremely high but finite curvature, avoiding the infinities of classical general relativity.

Loop quantum gravity has been applied to cosmology through loop quantum cosmology, which describes the very early universe. In this framework, the Big Bang singularity is replaced by a quantum bounce where a previous contracting universe reached minimum size and rebounded into expansion. This eliminates the initial singularity and may provide a natural mechanism for inflation.

However, loop quantum gravity also faces challenges. It has difficulty recovering general relativity in the appropriate limit. It does not naturally unify gravity with other forces as string theory does. It makes few testable predictions accessible to current experiments. The connection between the discrete quantum geometry at the Planck scale and the smooth spacetime we observe at larger scales remains unclear.

Other approaches to quantum gravity have been explored. Asymptotic safety proposes that quantum corrections to general relativity become weaker at high energies rather than stronger, allowing gravity to remain well defined at all energy scales. Causal dynamical triangulations build spacetime from simplicial building blocks and sum over all possible geometries. Emergent gravity suggests that spacetime and gravity emerge from more fundamental degrees of freedom, perhaps involving quantum entanglement or information theory.

Some physicists question whether quantum gravity is even necessary. Perhaps gravity remains fundamentally classical, never fully quantum mechanical, with quantum mechanics applying only to matter fields in a classical spacetime. Most physicists reject this idea because it leads to inconsistencies when quantum matter interacts with classical gravity. The majority view is that gravity must be quantized, we simply have not yet figured out how.

Experimental tests of quantum gravity are extraordinarily difficult because quantum gravitational effects become significant only at the Planck scale, where energies are about 10 to the 19 billion electron volts, far beyond any conceivable particle accelerator. The Large Hadron Collider reaches energies around 14 trillion electron volts, fantastically high by human standards but still 16 orders of magnitude below the Planck scale.

Some indirect tests have been proposed. Quantum gravity effects might slightly modify the dispersion of light traveling cosmic distances, with different wavelengths traveling at slightly different speeds. Observations of gamma ray bursts from distant galaxies could detect such effects. Gravitational wave observations might reveal quantum corrections to the waveforms from merging black holes. Cosmological observations of the cosmic microwave background might contain imprints of quantum gravitational effects from the Big Bang.

So far, no quantum gravity effects have been detected. This does not mean quantum gravity does not exist, only that its effects are too small to measure with current technology. Future experiments with more sensitive gravitational wave detectors, higher energy colliders, and more precise cosmological observations may eventually find evidence for quantum gravity.

Beyond quantum gravity, physicists seek explanations for other mysteries. Why does the Standard Model have three generations of fermions? Why are particle masses what they are? Why is there more matter than antimatter in the universe? What is dark matter? What is dark energy? Is there a theory that unifies all forces, including gravity, electromagnetism, and the strong and weak nuclear forces?

Grand unified theories, or GUTs, attempt to unify the electromagnetic, weak, and strong forces into a single force at very high energies. The idea is that at low energies, these forces appear different, but at energies around 10 to the 15 billion electron volts, they merge into one unified force. Evidence for unification comes from the running of coupling constants, which change with energy in ways that nearly converge at high energies, especially in supersymmetric extensions of the Standard Model.

GUTs predict proton decay, which has never been observed. The proton appears to be stable with a lifetime exceeding 10 to the 34 years, far longer than the age of the universe. Experiments in deep underground detectors watch huge volumes of water or other materials for rare proton decay events but have found none, ruling out simple GUT models and constraining more complex ones.

Supersymmetry posits a symmetry between bosons and fermions, with every particle having a supersymmetric partner differing by half a unit of spin. The electron has a supersymmetric partner called the selectron, the photon has a photino, the quark has a squark, and so forth. Supersymmetry could explain the hierarchy problem, why the Higgs boson mass is so much smaller than the Planck mass. It provides dark matter candidates like the lightest supersymmetric particle. It improves unification of forces at high energies.

Despite these theoretical attractions, supersymmetry has not been observed experimentally. The Large Hadron Collider has searched extensively for supersymmetric particles and found none, pushing the allowed masses of superpartners to increasingly high values. Either supersymmetry is broken at higher energy scales than expected, or it does not exist in nature. The lack of evidence has dampened enthusiasm for supersymmetry, though it remains possible that superpartners are just beyond current experimental reach.

Extra dimensions provide another avenue for unification. If space has more than three dimensions, gravity might propagate through all dimensions while Standard Model forces are confined to a three dimensional brane embedded in higher dimensional space. This could explain why gravity is so much weaker than other forces: gravitational field lines spread into extra dimensions while electromagnetic and nuclear forces do not.

Large extra dimensions or warped extra dimensions could be detectable through missing energy in collider experiments, where particles escape into extra dimensions, or through modifications to gravitational effects at short distances. Experiments have searched for such signatures without success, constraining extra dimension sizes and geometries but not ruling them out entirely.

The multiverse hypothesis suggests that our universe is one of many, perhaps infinitely many, each with different physical laws and constants. This idea emerges from several sources. Eternal inflation suggests that inflation never entirely ends but continues in some regions, constantly creating new bubble universes. String theory's landscape of vacua suggests different universes could settle into different vacuum states with different effective physical laws. The many worlds interpretation of quantum mechanics posits that all quantum measurement outcomes occur in different branches of reality.

If the multiverse exists, it could explain the fine tuning of physical constants necessary for life. In universes with different constants, stars might not form, atoms might not be stable, or chemistry might not support complexity. We observe a life friendly universe because we exist in one of the rare universes where constants allow life. This anthropic explanation is controversial, as it abandons the goal of deriving physical constants from fundamental theory, instead accepting them as environmental accidents.

Critics argue that the multiverse is not science because it makes no testable predictions. If every possible universe exists somewhere, any observation is compatible with the multiverse hypothesis. Proponents counter that the multiverse arises naturally from theories developed for other reasons, like inflation and string theory, and that statistical predictions about what typical observers should see can be tested.

The ultimate goal remains a theory of everything, a single unified framework explaining all forces, particles, and interactions, deriving all physical constants from first principles, and applicable at all energy scales from the Planck length to the size of the universe. Such a theory would represent the culmination of physics, the final answer to the question of what the universe is made of and how it works.

Whether this goal is achievable remains uncertain. Perhaps physics has fundamental limits, questions that cannot be answered by observation and experiment. Perhaps we will never fully understand quantum gravity or the Big Bang's initial conditions or the nature of dark matter and dark energy. Perhaps the universe is more complex and subtle than human minds can fully comprehend.

Yet the history of physics inspires optimism. Time and again, mysteries that seemed impenetrable were eventually solved. The nature of light, the structure of atoms, the origin of stellar energy, the expansion of the universe, all once seemed beyond human understanding. All were eventually explained through careful observation, clever experiments, and deep theoretical insights. The current mysteries may seem intractable, but so did previous mysteries before their resolution.

The quest for the theory of everything continues in laboratories and observatories around the world. Physicists build larger colliders, launch more sensitive satellites, construct more precise detectors, and develop more sophisticated theories. Each experiment constrains theories, ruling out some possibilities and narrowing the range of others. Each new observation provides clues about how nature works at its deepest levels.

We stand at an interesting moment in physics. The Standard Model and general relativity are extraordinarily successful, yet clearly incomplete. We have compelling evidence for phenomena they cannot explain: dark matter, dark energy, matter antimatter asymmetry, neutrino masses, the hierarchy problem, and quantum gravity. We have multiple theoretical candidates for physics beyond the Standard Model, but no experimental confirmation for any of them.

The next major discovery could come from anywhere. Perhaps dark matter will be detected directly in underground experiments. Perhaps the Large Hadron Collider or its successor will discover new particles or forces. Perhaps gravitational wave detectors will reveal quantum gravitational effects in black hole mergers. Perhaps cosmological observations will detect signatures of inflation or extra dimensions. Perhaps a purely theoretical breakthrough will unify quantum mechanics and gravity in an unexpected way.

Whatever the path forward, the goal is clear: to understand reality at its most fundamental level, to uncover the deepest laws governing existence, and to answer humanity's oldest questions about what the universe is and why it exists. This quest drives modern physics and represents one of the highest expressions of human curiosity and ingenuity. The theory of everything may remain elusive, but the search for it has already revealed wonders beyond imagination and will undoubtedly reveal more.

Conclusion. A Universe More Wonderful Than We Imagined

We have journeyed from the quantum strange to the cosmically vast, from the uncertainty principle to black holes, from the Big Bang to dark energy, encountering ideas that challenge every intuition and overturn common sense. Modern physics reveals a universe far stranger, far more subtle, and far more beautiful than anyone imagined before the twentieth century.

Particles exist in multiple places simultaneously until observed. Time flows at different rates depending on velocity and gravity. Space itself expands, carrying galaxies apart. Mass and energy are equivalent and interconvertible. The universe had a beginning and continues to evolve. Most of its contents remain mysterious, invisible dark matter and enigmatic dark energy dominating the cosmic inventory. These are not philosophical speculations but experimentally verified facts, conclusions forced on us by careful observation and measurement.

The revolution in physics is not merely a story of discovering new facts. It is a story of discovering that reality operates according to principles profoundly different from those our evolutionary history prepared us to expect. Our intuitions developed in a narrow range of speeds, scales, and energies where quantum effects are negligible and relativistic effects are tiny. When we probe beyond this familiar regime, into the very fast, very small, very massive, or very distant, our intuitions fail completely.

This failure of intuition does not represent a failure of understanding. Rather, it demonstrates the power of the scientific method. Through systematic observation, controlled experiment, and mathematical reasoning, humans can understand aspects of reality far beyond direct sensory experience. We cannot directly perceive quantum superposition, but we can design experiments to reveal it. We cannot directly experience relativistic time dilation, but we can measure it with atomic clocks. We cannot directly observe the Big Bang, but we can detect its afterglow in the cosmic microwave background.

The mathematical formalism of modern physics provides a language for describing phenomena that cannot be visualized or captured in everyday words. Schrödinger's equation describes how quantum wave functions evolve. The Einstein field equations relate spacetime curvature to mass energy distribution. The Friedmann equations govern cosmic expansion. These equations are not merely abstract symbols but precise descriptions of physical reality, allowing predictions that match observations to extraordinary accuracy.

This predictive success validates physics as a way of knowing. When quantum mechanics predicts interference patterns in electron double slit experiments, and experiments confirm those predictions, we gain confidence that the theory captures something true about reality. When general relativity predicts gravitational time dilation, and GPS satellites require relativistic corrections to function properly, we know the theory is not merely mathematical convenience but accurate description of nature. When Big Bang cosmology predicts primordial element abundances, and observations match predictions, we trust that the universe really did have a hot dense beginning.

Yet for all its success, physics remains incomplete. We have two magnificent theories, quantum mechanics and general relativity, that appear mutually incompatible. We observe dark matter and dark energy but do not know what they are. We do not understand why the universe has more matter than antimatter, or what caused inflation, or whether our universe is unique or part of a multiverse. These open questions drive current research and represent frontiers where human knowledge reaches its limits.

The incompleteness of physics should not diminish appreciation for how much we have learned. Three centuries ago, humans understood almost nothing about the fundamental nature of reality. We thought matter was continuous, time was absolute, space was Euclidean, and the universe was eternal. Every one of these beliefs has been overturned. Matter is made of discrete atoms and fundamental particles. Time is relative and can dilate. Space can curve. The universe had a beginning and continues to evolve.

This progress inspires both humility and wonder. Humility because it shows how much we once did not know and how much we still do not know. Wonder because it reveals that reality is far richer and more intricate than naive appearances suggest. The universe is not a simple machine following obvious rules. It is a quantum relativistic cosmos operating according to subtle mathematical principles discovered only through centuries of careful investigation.

Modern physics also reveals deep connections between apparently disparate phenomena. Electricity and magnetism are unified into electromagnetism. Electromagnetism and the weak nuclear force are unified into the electroweak force. Space and time are unified into spacetime. Mass and energy are equivalent. Waves and particles are complementary aspects of quantum entities. These unifications show that the universe is more integrated and coherent than superficial appearances suggest.

Perhaps the most profound lesson from modern physics is that reality does not conform to human expectations or preferences. Nature operates according to its own logic, indifferent to whether we find it intuitive or comfortable. Quantum mechanics is weird. Relativity is counterintuitive. The vastness of cosmic space and time dwarfs human significance. Yet these are simply facts about the universe we inhabit, and we must adjust our understanding to match reality rather than expecting reality to match our preconceptions.

This adjustment from intuition to evidence based understanding represents intellectual maturity. Children believe their direct perceptions reveal reality as it is. Adults learn that perceptions can mislead, that the Earth is round despite appearing flat, that objects fall at the same rate despite appearing otherwise, that the Sun is stationary while Earth moves despite subjective experience of the opposite. Modern physics extends this lesson further. Not only are perceptions misleading, but even careful reasoning based on everyday experience can fail when applied to unfamiliar regimes.

The antidote to misleading intuitions is mathematics and experiment. Mathematics provides a framework for reasoning about phenomena beyond visualization. Experiment provides empirical constraints on theories, forcing them to match observed reality regardless of whether they match expectations. Together, mathematics and experiment allow humans to understand the quantum world, relativistic spacetime, and cosmic evolution despite these domains lying far beyond direct sensory experience.

For students and readers encountering modern physics for the first time, the key insight is that reality is knowable but strange. You can understand quantum mechanics and relativity, not by visualizing them in familiar terms, but by accepting the mathematics and experimental results. The double slit experiment really does show particles existing in superposition. Muon decay really does demonstrate time dilation. The cosmic microwave background really does confirm the Big Bang. These phenomena may violate intuition, but they are as real as any everyday experience.

For society, modern physics provides both practical benefits and philosophical insights. Practical benefits include all of modern technology from electronics to medical imaging to nuclear power. Philosophical insights include understanding our place in the cosmos, the nature of time and space, and the fundamental constituents of reality. Both aspects matter. Technology improves lives. Understanding enriches them.

Looking forward, physics will continue to advance through a combination of better experiments and deeper theories. Next generation particle colliders may discover new particles and forces. Gravitational wave detectors may reveal quantum effects in black hole mergers. Cosmological surveys may detect signatures of inflation or extra dimensions. Theoretical breakthroughs may unify quantum mechanics and gravity. Each advance will refine our understanding and raise new questions.

The end of physics is nowhere in sight. Every answered question reveals new mysteries. The more we learn, the more we realize how much remains unknown. This open ended quest for understanding is what makes physics exciting. There will always be frontiers, always be puzzles, always be opportunities for discovery.

Living in a quantum relativistic universe means accepting that reality is not what common sense suggests. It means embracing the strange and counterintuitive when evidence demands it. It means understanding that the universe operates according to deep mathematical principles that can be discovered through reason and experiment. And it means appreciating that we are privileged to live in an age when humans have uncovered some of nature's deepest secrets while recognizing that many secrets remain.

The universe is more wonderful than we imagined, stranger than we expected, and more knowable than we hoped. Modern physics has revealed this wonderful, strange, knowable universe, and the journey of discovery continues. Every generation contributes new insights, building on foundations laid by previous generations, and leaving new mysteries for future generations to solve.

You are part of this ongoing story. Whether you pursue physics professionally or simply appreciate it as an educated citizen, you participate in humanity's quest to understand physical reality. The quantum world, relativistic spacetime, and expanding cosmos are not abstract concepts relevant only to specialists. They are the universe you inhabit, the foundation of the technology you use, and the context for your existence.

Reality is not what you think. It is quantum. It is relativistic. It is cosmic. It began in a Big Bang. It is mostly dark matter and dark energy. It operates according to precise mathematical laws that defy everyday intuition yet enable precise prediction and technological application. This is the universe revealed by modern physics, and understanding it represents one of humanity's greatest intellectual achievements.

The story of modern physics is ultimately a story about human curiosity and ingenuity. For thousands of years, people wondered about the nature of matter, the structure of the cosmos, the flow of time. Through patient observation, clever experiments, and deep reasoning, we have answered many of these questions and raised new ones. The quest continues, driven by the same curiosity that inspired our ancestors to look at the night sky and wonder what stars are made of.

We now know what stars are made of. We know how they shine, how they form, how they die. We know the universe had a beginning, what happened in its first moments, and how it evolved to its present state. We know that matter is made of quarks and leptons, that forces arise from gauge symmetries, that spacetime is dynamical and curved by mass energy. We know these things not through revelation or speculation but through hard won scientific investigation.

Yet knowing these things does not diminish wonder. If anything, understanding deepens appreciation. A sunset is no less beautiful for knowing that red light scatters less than blue. The stars are no less majestic for knowing they are distant suns powered by nuclear fusion. The universe is no less awe inspiring for understanding that it began 13.8 billion years ago and expanded from a hot dense state.

Reality is not what you think, but what physics reveals it to be: a quantum relativistic cosmos of extraordinary richness and beauty, governed by elegant mathematical principles, knowable through reason and observation, and far more wonderful than any fiction. This is the universe we inhabit. This is the reality uncovered by modern physics. And this is just the beginning of the story.

www.ingramcontent.com/pod-product-compliance
Lightning Source LLC
LaVergne TN
LVHW090417210926
848480LV00002B/598